T0282813

READY TO DIVE

Five Decades of Adventure
in the Abyss

READY TO DIVE

Five Decades of Adventure in the Abyss

Curt Newport

Purdue University Press · West Lafayette, Indiana

Cataloging-in-Publication Data is on file with the Library of Congress.

978-1-61249-966-6 (hardback)

978-1-61249-967-3 (epub)

978-1-61249-968-0 (epdf)

Cover image: *Greyhound aircraft*. A C-2A Greyhound aircraft, call sign Password 33, as it is being hauled out of the Philippine Sea by a Navy salvage team in May of 2019. I led the search team that initially located the plane's acoustic beacon at 18,500 feet, which resulted in further survey by Paul Allen's RV *Petrel* and the Vulcan team.

Contents

Preface

I hate the wind. I always have. If you're subjected to it for a period of hours, days, weeks, months . . . it just sucks the life out of you. It's even worse when it's raining. And when you're on board a ship, everything blows across the deck. The pages of your notebook flip, and your notes turn to rivers of blue ink. It tugs at your clothing continuously, like an invisible claw. I despise the wind.

Those who go out on the water on weekends in their plastic speedboats and sailboats don't know what it's like when you *have* to be at sea. They have a choice. "Oh, it's too rough. Maybe another day." Not so for us who work at sea for a living.

The winds are blowing 30 knots and there's a foot of seawater on deck? Sorry, put on your water-filled boots and get to work. Everything is dripping wet and you need to repair a 3,000-volt connector? Sorry, rig a tarp and try not to get electrocuted. It's 104 degrees in the shade and you've been forced to wear long heavy coveralls, gloves, and a brimmed hardhat? Sorry, hydrate and get to work. You drank too much the night before and your head is pounding like a jackhammer? Sorry, take some aspirin and stand by for a drug and alcohol test.

When you're at sea, the ocean environment is your enemy, one that must be conquered at any cost to get the job done. Of course, it's not always like this. There are times when the sea is flat, the sun is shining, it's not too hot or too cold, and the ship is handling well. Most days, the reality lies somewhere between those extremes. So why do people do it? I hope the words I've written will answer this question.

Charles A. Lindbergh was a hero to me while growing up. Here was a man who rose from airmail pilot to legend, successfully completing a single flight that changed the course of modern aviation. Lindbergh did much more, of course, such as scouting the South American routes for Pan American Airways. He was also the first international figure to be "canceled" by the media and government for his role as head of the America First Party and isolationist views.

Lindbergh wrote a wonderful analogy of memories. I cannot remember exactly which book that was in, but I have read everything he wrote. It was probably his *Autobiography of Values*, first published in 1976. In it he said something like the following: Memories are like a darkened room into which you can shine a flashlight. Some things are illuminated, while others remain hidden.

What that means is that memories are selective. For some reason, you remember certain things while others are destined to reside in a mental trash bin. The accounts of my life and various undersea operations have come from several sources: personal logbooks detailing the events of specific missions; company job reports and media collected during the job; official US government and military mishap reports; and my simple recollection of what happened and when.

I tried to keep a written daily log on every mission. Unfortunately, this was not always possible. Sometimes it was due to the frantic tempo of the job at hand; other times, I was simply too exhausted after a shift to write. These handwritten logbooks of varying sizes and styles now fill the bookcase in my office like battered relics from the past. One thing is certain: I will never miss the torturing rumblings of a bow thruster, the clanging sound of a ship's anchor chain, or a bunk vibrating over the top of the machinery spaces. Even now, when I hear a diesel generator starting up, I have a Pavlovian reaction and instantly feel tired. What I do miss, though, is the camaraderie of the men I worked with, as you develop close personal bonds at sea. Going to sea is a lot like going to war (except you're not being shot at). We all have each other's backs.

While no one is perfect, I have created as accurate an account as possible when describing the events and people included within the pages that follow. Buckle your seat belts and put your tray tables in the upright and locked position; it's going to be a bumpy ride . . .

C. Newport
Mt. Vernon, Virginia
January 2024

1

1978: A Near Miss with Nekton Alpha at 500 Feet

My eyes strained to see through the fogged viewpoint into the darkness in front. Was that another buoy line? I couldn't be sure. The steel hull of the two-man submarine swung slowly on a bubble of air in the Gulf of Mexico currents.

An occasional scraping sound reverberated through the hull as the vehicle rubbed against the concrete-coated pipeline.

"Why are we stopping?" asked a bare-chested Dave, a clueless diving supervisor curled up at my feet. Clueless because he was oblivious to the danger of being held to the bottom by the yellow poly lines. Dave was what they called at the time a "Chino hand" because he had learned to dive in the occupational commercial diving program at the Chino Correctional Facility. He was an easygoing guy incarcerated for five years for armed robbery. "Prison is the worst," he once told me. "Nothing worse than doing time and being locked up." He was more agreeable than Chico, one of his fellow former inmates. With his muscular upper body covered in jailhouse tattoos, his red bandanna, and Fu Manchu mustache, Chico was no one to trifle with, especially when he was drunk. Rumor had it he had knifed more than one person in the joint.

"I'm trying to find a path around that other line, and I can't maneuver well," I replied in exasperation. The reason I couldn't maneuver well was that, unbeknownst to my passenger, I already had one line in the sub's propeller.

I looked down at the high-pressure air gauge—1,800 pounds, certainly good enough to get back up. Seeing that we needed more oxygen, I cranked open a small flowmeter behind me. The balance of our breathable atmosphere depended on one large gauge, a war surplus altimeter pulled from a B-25 bomber. Depending on what it told me, I could either run a squirrel cage fan from a Chevrolet to filter carbon dioxide from our air or open the flow valve. I did the latter and the reassuring hissing sound increased.

The Nekton class two-man submarine I was piloting was technically not built in a garage, but by a skilled machinist named Doug Privitt in Southern California. With golf cart motors for power, a rudimentary life-support system, slide projector underwater lights, and two scuba tanks for high-pressure air, it was not sophisticated. The sonar was hand-cranked above my head, and it also used a standard aircraft gyrocompass for navigation—everything was by hand-turned valves, levers, and glass-faced analog gauges. As opposed to an acoustic tracking system, the vehicle trailed a small surface float. There was also a small telescoping steel grabber arm, which penetrated the sub's hull through a billiard ball. While it could dive to 1,000 feet underwater, you had to make it back up to tell the story, and at the moment I was concerned about doing that. Of course the Nekton was painted yellow, making it the proverbial Yellow Submarine. The thing could be towed on a trailer and weighed about 2 tons.

A bead of sweat dripped down my back as I stripped off my T-shirt and began wiping all the viewports. The single yellow line remained wrapped around the single propeller shaft. It still rotated, so at least I could move.

An underwater telephone known as the Gertrude crackled over the speaker as John O'Donnell asked, "How's it going? We haven't heard from you in a while."

"Well, I've got one line in my prop. I just didn't see it in time to avoid it."

"You still have maneuvering?"

"Yes, for now, but I've got another line to avoid," I replied. I could envision his boss, Rick, chain-smoking his Marlboro Reds as he paced the deck and cursed my ass.

Unlike these days of GPS satellites, in 1978 in the Gulf of Mexico, lay barges installed their pipelines by following a long breadcrumb trail of bamboo pole flags towering above small chunks of Styrofoam. The deployment of the floats was usually done before the job by a survey company using a high frequency radio mini-ranger system. The floats created a path for the barge to follow and were anchored to the bottom with poly line by small but heavy sections of railroad track. For us, the problem was that because the pipe was run out of the end of the barge's stinger, the poly lines would get caught under the pipeline, creating a massive submarine trap made of loops of poly line. It was a hazard that no one these days would think of diving in until the lines were cleared. But this was 1978, and in the Gulf, no one even gave it a second thought.

My simple task had been to inspect the pipeline for cracks in the concrete covering, and so far I had made a mess of it. This was also my first time as pilot in command of the Nekton, which made it even worse.

"Dave, stick your head up in the forward conning tower. Can you see that line trailing off to the right?"

"Yeah, I got it. Are we having problems?"

"Not yet," I replied in a less than convincing tone. "I'm going to try to slide past it down current. Keep an eye on it and let me know if we clear it."

Sliding the power lever forward, I gave the propeller some turns, cranked the rudder over, and for a moment it looked like we were going to make it OK. But then silt billowed out from the bottom and we were quickly in zero visibility.

We lurched to a halt, almost like we had run into a wall of Jell-O. We waited.

Eventually the dust settled, and it looked as though, somehow, one prop line had caught the other. Now I had a line around my conning tower as well. *Shit.*

My pulse quickened, not to panic mode, but heading in that direction. I could feel beads of sweat forming on my forehead and my face felt flushed and clammy.

"What do you think?" Dave asked.

"I think the dive is over."

My mind went into free-thinking mode as I assessed the situation. *OK, the Subsea International dive crew topside can find us because we have the surface float, if it is still attached . . . Can they even reach the bottom with their diving bell? If they could, why are we even here? . . .* I craned my face upward, but there were no viewports in the hatch. *What's the breaking strain of ¼-inch poly line? I should remember this from diving school. Is it 400 pounds, or closer to 700? I really need to remember this . . . How much volume is in the forward and aft ballast tanks combined? If I empty the high-pressure cylinders, will it give us enough upthrust to break the lines, or do we have to cut them? How much oxygen do we have left? Did we check the spare bottle before diving? I hope so.*

I also remembered an incident with a Nekton submarine eight years earlier in 1970. Nekton Alpha (the sub I was in) and Nekton Beta (the sub I was trained on) had both been attempting to raise a cabin cruiser from 230 feet of water. After the lift lines were attached, Nekton Alpha surfaced. For some reason, the pilot of Nekton Beta, R. A. Slater, decided to stay on the bottom. It was a costly error. Fifty feet off the bottom, the rigging parted and the boat fell back to the bottom, striking one of the conning tower viewports on Beta with obvious results: the sub started to quickly flood with seawater. The pilot somehow managed to get out of the thing and make a free ascent to the surface. The observer, L. A. Headlee, was not as lucky. He drowned.

I looked down at Dave and thought, *Sorry, buddy, but if we get stuck down here, you're going to die.* In reality, if we couldn't make it back to the surface and our surface float was missing, we were both going to die. A free ascent from 230 feet? Maybe. From 500 feet? No fucking way.

"Dave, did we attach the knife to the gripper?"

"Nope," came the disappointing reply in a shaky voice.

How much did that lead drop weight weigh? Wasn't it about 500 pounds? That would help, if we really needed it. I could crank the bolts and drop the propulsion motor. That wouldn't go over well, but it would clear the prop line . . . We're talking about not being stuck down here—that is the main thing.

The Gertrude crackled again. "How's it going?"

"I'm aborting the dive; we're coming up," I answered.

"Do you think the lines can hold you down?"

"I don't think so, but there's only one way to find out."

I poked Dave with my foot, telling him that the dive was over and we were surfacing.

I prayed that the inertia from the rising mass of the sub would make the lines break. Also, maybe reversing the prop would unwind that line, leaving only the one around the conning tower. But were the ends anchored to weights, or were they underneath the heavy pipeline? Either way, they needed to fail in tension.

I cracked the high-pressure valves open as the hiss of air filled the interior of our steel tube. Mindful of inadvertently freezing the valves, I slowed the rate a touch, watching the tank gauge like a hawk. Slowly, bow first, we started rising as the bubbles of air in the tanks swelled in volume.

As the Nekton picked up momentum toward the light, I reversed the propeller and observed with joy as one of the lines jumped out of the tail and floated away. At 100 feet from the bottom, the conning tower line momentarily snagged on the compass stand, then miraculously slid clear. We were free!

The Nekton's conning tower burst through the warm bubbling Gulf waters as the sub bobbed up and down like a steel cork. I could see our ship, the *Antares*, through the porthole as John and Rick observed our surfacing. Brazenly, I cracked the pressure equalization valve, pushed the hatch open, and stood up on the pilot's stool. I grabbed a pack of Marlboro Lights from my pocket and lit up, taking a long, deep drag of nicotine. As the ship slowly maneuvered toward our location, John pointed and laughed, while Rick gave me daggers. I gave them both my best shit-eating grin and waved. It was over and I had dodged a bullet again.

2

1951–1966: Army Brat

was an Army brat. During most of our trips from one posting to another, I would lie down in the back window of our car while Dad drove us cross-country. I don't know whether I simply enjoyed watching the stars or liked the comfortable position under the rear window. It didn't matter whether it was our Chevrolet Bel Air or the old Buick with the circular vents on the front fenders. For some reason I enjoyed it. Then there was the grand tour of Europe in 1960 with my brother and me stuck in the tiny back seats of a 1960 Austin Healey 3000, taking secret 8 mm film footage of East Berlin among the Russian troops (before the wall was built). That one experience showed me the difference between freedom and Communism. In West Berlin, there was activity and life. In East Berlin, there was nothing but darkness and bombed-out buildings left over from the Second World War. No one there looked very happy. It made you think.

We were always moving. By the time I was sixteen years old, I had not lived in any one place longer than a year and a half by my memory. Dad, Mom, my older brother, Chris, and I were always packing up and moving somewhere. From the Presidio in California; to Enterprise, Alabama; to Poitiers, France; Mannheim, Germany; Arlington, Virginia; Ft. Leavenworth, Kansas; Atlanta Army Depot, Georgia; Florissant, Missouri; Oakland, California (while Dad was on his first tour in Vietnam); and then finally to Vienna, Virginia. But on this voyage in 1967, I was not happy at all. Why? Because I had been forced to leave all my long-haired, desert-boots-wearing, pot-smoking friends in California and head to the East Coast to Virginia, toward what appeared to be certain doom.

Dad had already whacked off all my hair earlier during his one brief trip back to Oakland from Nam, leaving me unrecognizable to my buddies at Bret Harte Junior High School. In a handful of days he had come home, dragged my ass to the barber shop, and saddled me with "white sidewalls." I walked like a stranger among my friends, who didn't even recognize me. It was as though I had been branded. No more looking cool; now I looked like a square. That was only the beginning, because before we made our stop at Dad's brother Norsuda Newport's house in Ohio on our way to Virginia, I was once again put in my place in another anonymous barber shop. After riding in the back of our Chevy for days, when we pulled into that strip mall and I saw the sign, I knew what was coming. Not as bad an attack on my person, but an attack, nonetheless. Being subjected to such a personal affront made me hate my father at the time.

Even with that, Chris and I had a good time with Norsuda. He was also a US Army officer (retired), and he was a medical doctor. (It was only decades later that I learned that he was not my dad's brother, but his uncle.) Norsuda was a bit more laid-back than my dad and had a small pontoon boat that he let me drive around a small lake, much to my mother's distress. He let Chris and me have fun and do stuff. He taught us things. He gave us knives he had picked up from Indians in Panama. He let us play with guns. While visiting us in Alabama, he showed us plants with leaves that closed at the touch of a finger. Norsuda was cool.

By that time, after much instruction at the Army range, Chris and I were proficient with both pistols and rifles. By age thirteen, I could field-strip and reassemble an Army-issue Colt 1911 .45 caliber automatic blindfolded. But did I take it to school and shoot up the place? No way. Things were different then.

But long before that trip from the land of long hair and free love to the straitlaced East Coast, I had my beginnings in 1951 at Peralta Hospital in Oakland, California. I suppose you could say that my first act in life was to almost take out my mother while being born. According to my father, I was too long for my mother's womb and almost killed her while she was giving birth. (Upon my entry into the world I had crooked feet, and although for some time the doctors were not sure I would walk, they fitted me with small braces, and over time my feet straightened out to function as useful append-ages.) Mom wore the scar of my existence for the rest of her life. It was not a good childbirth experience for Flora Arakelian-Newport.

I was the second son of the family, with an older brother, Christopher Mark Newport, a mother of Armenian descent, Flora, and a father, Elswick, with a shadowy background of sorts. We never really knew much about Dad while growing up, but it was obvious that he had had a difficult upbringing, starting in Oklahoma. Over time, we learned more about it, especially the fact that he was raised by his grandparents, who gave him their last name, Newport. His last name would have been Abshire because that was who his father was. But he never knew his dad, only his grandparents and mother, Bernadine. In fact, I didn't even know until I was seventeen that my father had been briefly mar-ried before meeting Mom and that I had a half-sister.

My uncles on my mother's side (except for Bobby) never married, choosing instead to run their produce business on Fourth Street in Oakland. Charlie was a character. With his mustache and the fedora he always wore, he could have passed for Howard Hughes. Charlie used to lecture me about the econ-omy. "Ya know, jobs are hard to come by these days . . . ," he would say. I never reminded him that the Depression was long over. He'd also come up with lit-tle gems during an especially damp and overcast Bay Area day; for example: "This would be a good day to be in the sack with a blond."

Charlie was a skilled mechanic and worked for Mr. Boeing before Boeing Aircraft even existed. He served out the war repairing Navy aircraft in Oakland, later explaining to me the intricacies of sliding a fabric covering over the wing structure of a Stearman and telling me about people calling him nuts when he once suggested that aircraft would eventually have retractable landing gear. He also invented and fabricated a production line for the produce business that could have made all of them into the new Giant Foods or Safeway because they were packaging and selling vegetables in plastic bags locally before anyone. But they didn't care. They had Arakelian Brothers and that was the end of it.

George was not quite as open as Charlie, but he ran the place with a steel fist. His most remarkable features were a large birthmark on one cheek and his ability to pull fourteen-hour days, seven days a week, for forty years.

None of them had a social life. They got up at 4:00 a.m., drove down to the shop, made their deliveries of carrot sticks, mixed vegetables, and spinach, came home, and by 7:00 p.m. they were all snoring away in front of the tiny black-and-white television in my grandmother's dining room. I don't think they ever drank or smoked their whole lives.

Out of all of them, George was the only one to experience combat during the war, managing to somehow bring back a 9 mm German Luger, a helmet, and a large leather belt, complete with the German Army belt buckle. George talked about the war only once after I had returned from a tour of the invasion beaches in Normandy.

"Did you get to Sainte Mère Église?" he asked me.

"Why?"

"Because I was there . . . Third wave, Omaha Beach, with the Army infantry."

That was all he ever said. But I knew he didn't get that Luger playing around at the rear. Like most of the Greatest Generation, he just wanted to forget it ever happened. Then I remembered the yellowed fragile newspaper clipping under the glass top in my grandmother's bedroom. It showed George getting a big hug and kiss at the train station from his mother, Lucy, happy to see that her son was still among the living. He was one of the lucky ones.

Johnny was a cigar-chomping hard-ass who hated kids, especially my brother and me when we visited the house at 591 Capell Street. He was shorter than his brothers and always giving us grief for running around, foraging through the basement crawl space for WWII artifacts, setting fire to the garbage in the kitchen, and causing general mayhem. But as opposed to his three brothers, Johnny liked to party, as in have a life. He worked for decades as a foreman at a nearby General Motors Plant and always drove Chevys, whether it be the 1952 Bel Air or the 1962 Impala. Charlie liked Pontiacs. His 1948 convertible with a straight-eight-cylinder engine sat in their driveway for probably thirty years after he lost the keys, the battery went dead, and the engine froze up.

Johnny enjoyed frequenting the strip clubs, cabarets, or whatever they were called then in San Francisco. You could always tell when he was heading out for a night on the town—him all dressed up in a sport coat, slacks, and tie. I only saw one picture of him with a girlfriend, a tall redhead who, not surprisingly, looked like a retired stripper.

While Bobby worked at Arakelian Brothers alongside George and Charlie, he wanted to have a life outside of work and was married for many years. At least when I was young, Bobby had a mean temper and always seemed angry about something. Maybe he was sick of wasting his life at the spinach shop, or maybe he was just angry in general. I never knew.

Bobby served in Korea with mixed success. In fact, when my father was flying Bell H-13 helicopters for the 4077th MASH unit during the war, they found a way to meet up. I don't think Bobby liked the Army much, especially when he was disciplined after getting separated from his unit in the field. There was also another story, about how he had been ordered to man a machine gun emplacement and refused. His superior officer did it instead and was reportedly killed in the process. Maybe that was what haunted Bobby.

I can recall bits and pieces of the various locations where we lived while Dad was hauled by the Army from one part of the world to another. My earliest memories are of living at the Presidio. I remember sitting in the back seat of a car while being driven home from a regular visit to my grandmother's house. Just off the jagged rocks, I could see the rusted mast of a ship jutting out of the water. I never knew what ship it was, but it was something I always looked for during the drive along the San Francisco coast on the way to our small duplex on the base. It was probably the remnants of a fishing boat. One day it was gone, probably torn to pieces by a storm.

Other thoughts remind me of the Presidio. Like the times when a young friend and I would jump up and down on the hoods of the cars parked in the garage. Or when, for some reason, we decided to strip off all our clothes and walk around naked. I think my mother made short work of that. But the most important memory of that time in my life was when I and probably that same young friend played on top of the parking garage, a brick structure topped with tar and gravel and lined with barbed wire. After a while, his mother found him and took him away from what was obviously a hazardous situation. But for some inexplicable reason, she left me there alone. Maybe she thought I was someone else's problem (which I was). To this day, I don't know how I did it, but I managed to fall off the top of the garage and land on my head, mostly on the right side of my face. (This might explain many things about me—dropped on my head as a youth!) It had to have been at least a 10-foot fall. I stumbled home and found my mother having a coffee party with some of the local Army wives. I tugged at her skirt in an attempt to get her attention, but she kept pushing me away. Then finally she looked at me, and seeing the swollen and bruised side of my face, shrieked in horror. She immediately took me down to the nearby dispensary. They didn't do much for me there, except spread some iodine on the area and pass along that my cheekbones had probably been fractured. As a result of that incident, my face is asymmetrical. One side looks a bit different from the other. Somewhere in the family archives of early photographs is an image of me with a very swollen face looking unhappy.

Decades later, Chris and I located the duplex where we had lived in the 1950s. He mentioned that our next-door neighbor had committed suicide by swallowing the barrel of his 1911 .45 caliber Army-issue automatic. I had no recollection of this because I was probably too young at the time.

As the day went on, we found the remnants of Crissy Field, the runway where our father had taken off piloting various Army aircraft. The runway had been torn up and replaced with a large expanse of grass, where the locals were flying kites. It was as if they were trying to remove all identification of the place as a former military base. But that is what it had been, at least when I lived there. This bothered me.

This was the place where I would sit in the cockpit of an Army L-20 Beaver and play with the control yoke while my dad planned his flight. Just a few miles down the beach was where my father had almost drowned after ditching his Ryan Navion (L-17) in the frigid waters of San Francisco Bay, just off Fort Mason. From the *San Francisco Chronicle*, August 3, 1951:

Army Light Plane Crashes in Bay; Pilot Unhurt: A single-seater Army liaison plane crashed in the bay off Aquatic Park yesterday afternoon and the pilot was rescued from the water minutes later.

The plane, an L-17 model, was piloted by Captain Elswick Newport, 6517 Eastlawn Street, Oakland, attached to the 7th Armored Division at Camp Roberts.

Captain Newport had brought some papers to the San Francisco Port of Embarkation from Camp Roberts. He took off from Crissy Field on his return trip at about 12:15 p.m. His plane plunged into the bay a few minutes later.

Authorities at Fort Mason saw the plane crash and dispatched three rescue boats.

The flier was in the water only five minutes before being rescued.

What the brief article didn't say was that my father had initially planned to set the aircraft down on the shore, but there were too many people out walking around, so he was forced to ditch. He wasn't even wearing a life vest and didn't have time to open the canopy of the plane before he glided it down into the cold waters of the bay. My father stood on the wing briefly, until the Navion started to sink, then jumped in the water and paddled away as he watched the aircraft submerge behind him. He was hauled out of the water by a Coast Guard Cutter and given a shot of the smoothest whiskey he had ever tasted, before being sent on his way. Dad had just been paid in cash, so the rest of the day was spent drying out his paycheck in the form of US currency on the clothesline.

From the Presidio, we were sent to Fort Rucker, Alabama, where Dad learned to fly helicopters. We lived off base there in Enterprise at 400 Doster Street, and one thing I can remember is that it was boiling hot. I was very young at the time and would often follow my brother to kindergarten and hang out in the back of his classroom because I had nothing else to do. I don't think Chris liked it very much because having his punk younger brother tag along was embarrassing to him. Our neighborhood was a new one with gravel roads, and I had no other children to play with. As a result, on several occasions, I thumbed through the telephone book, calling up total strangers to see if they had a youngster who could be my friend. While walking back from crashing Chris's kindergarten class, I would often stop off at a local dairy plant and buy a small carton of chocolate milk for five cents. Coca-Colas in small glass bottles from vending machines cost a dime.

We lived in Alabama for a year or so, then it was off to Europe, initially Poitiers, France. This was in the era before jet travel, and Dad left before the rest of us to set up some sort of household off base. It was Christmas 1958 when we departed La Guardia Airfield in a C-54 military air transport four-propeller radial-engine aircraft for what was to be a very long journey. In those days, there was absolutely nothing to do during what was a long and very noisy flight. (Unless you've spent more than ten hours on a four-engine propeller-driven aircraft, you cannot comprehend the noise level.) We were given a small Magic Slate cardboard drawing toy, small plastic wings to pin on, a tour of the cockpit, and that was it.

The thing I recall about this voyage was that it was dark the whole time. But that's impossible, isn't it? It was dark when we left New York, dark when we stopped in Iceland to refuel, dark when we landed at Orly Field, dark on the electric train to Paris, and dark when we finally got to the house in Poitiers. And it was snowing. My life experience being California and Alabama, I had never seen snow before. The cool thing was that Dad had the place all set up with furniture and a Christmas tree.

Despite the help their country received from the US Army during WWII, the French seemed to hate having us (Americans) in their country. The maid who cleaned our house regularly stole my mother's jewelry.

Every morning, I was hauled to the military base (in the dark, of course) for school, including French lessons (I never learned French or any other language). The rest of my time was spent playing in a trash dump near our house and accompanying my mother to the airfield, where she would wait for Dad to return from a flight under the leering eyes of other pilots.

Chris had some fun in France, especially during a Boy Scout outing on a beach somewhere on the southern coast. During the camping trip some of the boys discovered a barnacle-encrusted German mine washed up among the seaweed. The trip ended when the US Army Explosive Ordnance and Disposal team was called in to blow it up.

After only a little over nine months in France, we packed up the house once again and headed off to Coleman Barracks in Mannheim, Germany. Unlike in France, we lived on base in a good-sized three-bedroom apartment. Like all the other installations, Coleman had an airfield, so Capt. Newport had a place to fly.

The difference between the Germans and the French was striking to my sensibilities as an eight-year-old. Even fourteen years after WWII and the influx of money from the Marshall Plan, Germany was still a defeated country. And the local populace seemed to respect us as the conquering Army and appreciate the positive effect having us there had on the local economy.

We had a small black-and-white television while there with maybe two or three channels. German television at the time didn't have commercial breaks; they were all shown at the end of the regular programming, to my recollection. At either the beginning or end of each day's broadcasts, there was footage of a local Bürgermeister (mayor) blowing a small horn to call out the deer or something.

When I wasn't in school, many of my days were spent wandering through the dense forest either on or near the base. I don't know whether it was a difference in resource management or just the nature of the Black Forest, but I recall very little underbrush, just tall trees with a high canopy overhead. It was in places like this that Chris and I would explore the numerous concrete German bunkers from the war, searching for "souvenirs." We never found any, but what we did notice were patterns of large overgrown holes scattered haphazardly throughout the forest. They were huge, at least 30 feet in diameter and as much as 10 feet deep. We learned that they were bomb craters from WWII, created by the thousands of B-17 and Lancaster bombers that laid waste to the whole country.

The officers at Coleman Barracks were always trying to come up with "activities" for the dependents, mostly to get us out of our parents' hair for a while. One time someone decided to make us all go on an Easter egg hunt on the base. I don't know who the hapless NCO was that had to organize this detail, but he made it a memorable one, at least for me.

First, we were all collected up on several Army buses as though we were going to the front lines. After thirty minutes of travel on bumpy dirt roads, we were unceremoniously dumped out in the middle of nowhere, probably on the fringes of the base, near a forest and a large barren clearing. It seemed to be an odd place to hunt for Easter eggs, but we all took it in stride and started foraging through the underbrush and peculiar dirt pits.

So here I am, eight years old, up to my elbows in dirt digging around for these Easter eggs in this long, deep dirt pit. After a while, I started finding strange-looking sharp fragments of metal and other dirt-encrusted components. I cleaned off one of them and, to my surprise, found myself holding the fuse from an exploded hand grenade. (As an Army brat, I knew what fragmentation grenades looked like.) It wasn't long before I realized that all of the irregularly shaped parts were shrapnel from grenades. *Oh joy! Hand grenades!* For the uninitiated, what I was finding were the razor-sharp fragments of the standard Army Mark 2 grenade, which typically carried an explosive charge of about 52 grams of TNT. *Only* the US Army could have thought that the location where they practiced throwing hand grenades was the ideal spot for an Easter egg hunt. Yep, that NCO got the officer back good. You want an Easter egg hunt? I'll give you an Easter egg hunt!

I was just about to announce my great discovery to my brother when I noticed that everyone else was gone. I had been left behind. *What?* They didn't even do a head count before they left? My brother didn't bother to make sure I was there? These days, this would be the impetus for an excellent lawsuit.

Not knowing what else to do, I started walking back in the direction that I *thought* we came from. I was getting scared and the inevitable crying started, with tears streaming down my cheeks. (Cut me some slack. I was not even ten years old and was stuck in a scary forest and it was starting to get dark and cold. At least that makes it a better story, anyway.)

After about thirty minutes a Jeep came along carrying two GIs. They stopped and asked what in the hell I was doing all alone in the fucking forest. I explained what happened and they just thought it was the funniest thing they'd ever heard. "Har har har ... Dumb kid got left behind ... Har har ... Wait till the CO finds out about this one!" I was less than amused, but they were in stitches. I told them where I lived and we headed off, bouncing down the road. They used the Jeep's radio to tell the base about their hilarious find and I could hear a lot of laughing on the other end. Yeah. So funny. The guy in the passenger seat kept trying to get me to say something to their commanding officer. As scared as I was, I shook my head no and they gave up after a few tries.

Finally, we made it to the front of our apartment building near the dispensary. I thanked them and they drove off (still laughing of course), then I headed up to the third floor ready to beat the hell out of my brother ... Well, not really. He was bigger than me.

So that was an Army Easter at Coleman Barracks, Manheim, Germany, in 1959. I don't recommend it.

As another activity, the Army decided to cram us into an olive-drab bus and take us to "summer camp," high up in the Black Forest. Once we arrived, I felt as though I had been drafted and sent to the front lines during the Battle of the Bulge. The camp was laid out just like a bivouac, with a line of standard-issue tents with the words "US Army" stenciled on the outside and an open area where a large canvas water bag was suspended by a tripod. All that was missing was a stack of M1 Garand rifles and some artillery pieces. Inside the tents were rows of cots. They didn't look very comfortable, but of course, we were in the Army now.

I cannot remember whether our camp counselors were civilian or military, but they were older than all of us, who ranged in age from maybe eight to ten years old. (There were no girls, just us recruits.) We just fooled around the first day—in nearby bomb craters of course—until it was almost time to hit the sack. It was starting to get cold and very dark when they lined us all up in front of the tents to outline the next day's activities.

As we stood shivering in the twilight, our "section leader" thumbed through his clipboard. Within a minute or two, his assistant briskly walked out and handed him a piece of paper. As he looked down at it, his expression became stern. Then he read from the paper in a very serious tone: "We've just received word that the monster known as Red Eye has escaped from a laboratory and is reported to be in this area. This is an extremely dangerous creature so keep an eye out for it!"

I remember at the time not buying their ridiculous announcement, but some of the younger kids started whispering among themselves and looked scared. I didn't say anything but thought about it a little . . . *Did they ever catch that guy Mengele? Did he make it to South America like the other Nazis? The Germans were doing some real weird experiments during the war . . . Nah,* I concluded. But I must admit that as an eight-year-old well-versed in military history, I still felt a chill run through my skinny body.

During the night as we all tried to sleep, it was deathly quiet and pitch-black outside our tents. Sometime during the night, I woke up. Still half asleep, I heard a noise. *What was that?* I wondered. Then it started . . . a scratching sound. I bolted upright from underneath my two woolen blankets, listening intently. Scratch . . . scratch . . . scratch . . . , then a low growl.

I cautiously got up from my cot and made my way to the tent's front door flap, lifting it open. Then I saw it, a dark shadowy figure running through the camp, topped by a glowing red eye! Someone shouted, "It's Red Eye! Run for your lives!" followed by a bloodcurdling scream.

A wave of terror swept our tent as many of the kids hid under their blankets and curled into little balls of fear. A few of them sobbed, "I don't want to get eaten by Red Eye!" A couple of others yelled for their mothers. Many were openly crying and terrified. More scratching and growling ensued. Eventually it stopped, and all was quiet again.

The next morning, our counselors taunted us with their little smirks, satisfied that they had succeeded in scaring the hell out of a bunch of little kids who only wanted to enjoy the forest and some time away from their parents. That was my summer camp.

My father always told Chris and me that we were direct descendants of Capt. Christopher Newport, the famed privateer who helped found the Jamestown Settlement in Virginia. Officially he was known as Christopher Newport of Limehouse, Mariner. While he was best known as the master of the *Susan Constant*, the vessel that brought John Smith to Jamestown, he had enough exploits to fill anyone's lifetime:

> Christopher Newport, whose voyages across the Atlantic between 1606 and 1611 did much to secure the precarious foothold of the English on American soil, has always been a somewhat obscure figure. . . . Newport was born in 1560 and doubtless went to sea as a boy. The first mention of him does not occur until 1581. . . . English merchants were taking advantage of the situation created by the Spanish conquest of Portugal.

Newport fought in the Battle of Cadiz under Sir Francis Drake and was one of Drake's supporters in his quarrel with William Borough. During his first voyage to the West Indies, the thirty-year-old mariner took command of the *Little John*, one ship of a small fleet of three belonging to John Watts (the other two were the *John Evangelist* and the *Harry and John*). He eventually sighted fourteen Spanish merchantmen and managed to force two of the enemy ships aground on the Jamaican Coast. What happened next cost the young sea captain dearly:

> In the bitter struggle which ensued, five of his men were killed, and amongst the seventeen injured was Newport himself, who lost his right arm. . . . It was a heavy price to pay, for one of the ships sank before the captors could unload the silver, and the other, having run aground, was saved from the English . . . by the arrival of galleys from Havana.

On October 14, 1724, one John Newport was convicted of grand larceny during a proceeding at the Old Bailey in London. The proceedings read as follows:

John Newport, of the Parish of St. Brides, was indicted for feloniously stealing 4 s, the Money of George Dascomb, the 1st of this instant October. It appear'd by the Evidence, that the Prisoner was Servant to the Prosecutor, and
having several times lost Money out of his Breeches Pocket, he put in some remarkable, and that it was taken upon the Prisoner the next morning. The Jury
found him guilty of the indictment and he was sentenced to "Transportation."

In those days, "Transportation" meant deportation to either Botany Bay,
Australia, or the New World. Mr. John Newport, assuming he had a choice,
wisely chose the New World or, more specifically, a voyage from Newgate
Prison (their version of a supermax at the time) to Virginia, beginning on
December 14, 1724. The North Atlantic in December on a sailing ship was not
a place you wanted to be, to say the least.

All that for basically stealing twenty-five cents in today's money. Of course,
like all the prisoners at the fictional Shawshank prison, he was innocent, right?

The blood connection between this John Newport and Capt. Christopher
Newport of Limehouse is tenuous, but probable, according to contemporary
genealogists. The Newport sentenced on Saturday, October 14, 1724, at the
Old Bailey could have been a grandson, great-grandson, or nephew. Who
knows? But given that the famous sea captain was basically a pirate backed
by the British government, it is not a stretch to make the connection. Many
of the records from that time period were destroyed during the bombing of
London in the Second World War. However, there is no denying my connection to John Newport, "master thief." That, unfortunately, is well documented.

The trail of my existence in the United States began when John Newport
was transported from Newgate Prison and arrived in Orange County, Virginia, in 1724 as a prisoner. From that point in time, the Newport band of cut-
throat pirates bred like rabbits and spread like a swarm of locusts, raping and
pillaging their way southwest through Virginia, into Kentucky, down through
Tennessee, and eventually ending up in Oklahoma, where my father was born
in 1924. Well, not exactly.

John Newport, the first of my father's clan to make it to North America,
eventually became a deputy sheriff in Orange County, Virginia, according
to a period land patent book. He was also given a plot of land in 1740 to
make his own in the county "on brs. Of Black Walnut Run, bounded by John

Smith & Ivy Br." All I can say is, Who would make a better sheriff than a former criminal?

I believe his marriage to Harriet Burnett produced a son, Richard, born in 1744 in the city of St. Mark, Virginia. Unlike his father, Richard became a pastor—possibly to invalidate his father's sins.

From Virginia, the family expanded into Kentucky and then Tennessee, first with Ezekiel Newport (born in 1777 in Culpeper County, Virginia), then his son Joseph Cecil Newport (born about 1825 in the area of Morgan County, Tennessee), his son James Montraville Newport (born in 1844 in Morgan County), and his son Elswick Washington Newport (born in 1878 in Scott County, Tennessee). Elswick married Kathryn LaGrange and they settled in Oklahoma, where they raised my father's mother (Bernadine) and Uncle Norsuda, and then my father (their grandson). The trail of military service appears to have begun with Joseph, who served in the Civil War, with his son James serving in the Kentucky Volunteer Infantry, and James's son Elswick serving in the Spanish–American War.

My transition from an Army brat to a hippie did not go smoothly for me or my father. I was very rebellious and didn't like being molded into something I did not want to be. In reality, I was (and will always be) an Army brat, and in spite of my appearance I remained a patriot who loved and believed in America and the country's values. The problem was that, to my father, I didn't look the part. The times were changing, and the country was in turmoil, all because of the Vietnam War. While some of my friends later protested the war, I never did any such thing. In fact, even to this day, I have never protested anything. I have always felt that it is a waste of time and better to dedicate oneself to creating actual change, instead of spending time protesting. The reality was that despite our disagreements, I respected my father and his military service. Why would I not? It was really a superficial difference of opinion. He wanted me to look like the son of an Army officer, and I wanted to look like a Beatle. It all came to a head after our cross-country trip when Dad got back from his first Vietnam deployment as part of the Army Concept Team in Vietnam (ACTIV).

3

1967: Runaway

B y the time my father got back from his first Vietnam tour, he had it in for me in that he was going to "set my mind right," mostly by cutting off my hair (which really wasn't all that long) and forcing me to wear appropriate attire. Gone were the Beatle boots, confining pants, and paisley shirts. In short, my identity would be taken away from me and a new one substituted.

The deed was done during a cross-country trip from Oakland to Washington, DC, just before I was to be presented to Norsuda Newport, the man he called his brother who was his uncle.

After several days of the typical Newport family cross-country trip, we entered what I viewed as the squarest, most uncool part of the country I had ever seen: Northern Virginia. I cannot recall where we initially set up camp, probably a motel in Arlington, near the Seven Corners shopping center.

I can still recall my mother's approving comments as we trolled past the straightest-looking teenagers I had ever seen in our 1965 Chevrolet: "Look at them! See how they're dressed! Look, Curtis, their hair is so nice and neat . . ."

Dad had been promoted to full colonel and had a new job at the Pentagon, working for the assistant secretary of the Army, Dr. Ron Fox. It was a very high-level position, as I would later figure out. At the time, I couldn't have cared less. All I knew was that anyone in the world I could call a friend was 3,000 miles away and I was alone, had a new personal identity forced upon me, and was miserable.

It was different this time because I was becoming an adult. Before when we moved all the time, I accepted it. It went with the territory of being an Army kid. Certainly it was always difficult emotionally being the new student introduced to the class, not having any friends, or even knowing anything about where I was living. But I was able to handle it because it was all I knew. This time, though, I was rebelling inside. I asked myself, *Why did I have to always go through this? Why could we not live somewhere for more than eighteen months?* I hated always being lonely and having to leave my friends and being forced to make new ones. Just being a kid was bad enough, but this nomadic lifestyle just made it worse. By then I was sick of it.

The Newport tribe ended up in a large home in Lake Vale Estates in Vienna, Virginia, at the time considered the boondocks. Mom and Dad got the model home of the development for $48,000, a lot to pay in 1967 for a four-bedroom, two-and-a-half-bath house with a full basement and ample two-car garage. For the first time in my life, I had my own bedroom, sleeping in the same single bed I had in Enterprise, Alabama, in 1958. My brother, Chris, was not with us. He stayed behind in Oakland, where he was attending junior college.

James Madison High School was within walking distance of our new home, and it was not long before I was dutifully enrolled in the tenth grade. From the first moment I became acquainted with the school's administration, I knew there were going to be problems.

Walking down the halls between classes, I quickly decided it was best to be as invisible as possible to avoid confrontations. I usually hugged the walls of the hallways and walked very quickly with my head down. But that didn't stop the threats. Every day or so someone would point me out, pronouncing that they were going to give me a pounding after school. I think part of the problem was that they perceived me as part of the counterculture, not realizing that I came from an Army family and had a father who had just returned from combat in Vietnam. Had they known this, I think it might have been different. But they didn't, and nothing changed. The funny thing is that by today's standards, my outward appearance was not very radical. My hair was not even very long. Regardless, I didn't fit in and paid the price for it.

At some point, I simply decided that I had had enough. I was finished with the threats of violence from a bunch of moronic jocks, jokes about my clothing, and comments about how different I was from everyone else. In 1967, going from the Bay Area to a place like Vienna, Virginia, was like going to another planet. Culturally speaking, I was way ahead of the curve, and they were all stuck in an East Coast mentality in both how they thought, what rules they followed, and how they dressed. To that end, I hatched a plan to run away back to California and Nirvana that rivaled anything developed by SEAL Team Six. Or so I thought.

First, I needed an alibi that would allow me to make it to my destination before my parents knew I was really gone. To that end, I let them know that I was playing a band gig and would be spending the weekend at a friend's place. That would at least give me a running start.

I also needed a new identity. I forged an ID card with a photo, giving me the new name of Ross Spicer, an old grade-school friend from when we lived in Missouri.

I also knew the first thing my parents would do when they realized I had left was give my photo to the police. I reviewed all our family albums and collected up every image I could find of myself. They were going to have to look hard to find anything to give to the cops.

I also needed travel money and a way to get to the West Coast. I forged my father's signature on a check made out to cash in the amount of $125, enough to pay for a one-way plane ticket from Dulles Airport to San Francisco International and have money left over (plane tickets were a lot cheaper in 1967).

I packed some extra clothes in the case for my 1965 Fender Precision Bass as well as a small bag. The only other thing I planned to take was my 50-watt Fender Bassman amplifier head. I guess I must have figured I could make money playing music.

I arranged a ride to the airport with a school friend, cashing the check along the way at our local bank. I don't think I realized at the time the seriousness of what I was doing. Anything could happen. I mean, I had a support base there with all my high school friends, but where was I going to stay? How would my mother and father react? Would they try to bring me back, or let me stay? A lot of unknowns, but I didn't look back after boarding the Boeing 707.

Now, more than fifty years later, I can discern a few holes in my plan. Was I going to stay there permanently? What did I think—that some family was simply going to take me in and let me go back to Skyline High School? How was I going to support myself? Did I really think my parents were going to let me go?

All I knew was, I was not going to take it anymore and was getting the hell out of Dodge.

My friends welcomed me with open arms when I arrived in San Francisco that evening and spirited me off to a student's house, where he lived with his half-sister and a mother who spent most of her time traveling for business, giving them free rein of the place. Unfortunately for me, his mom was home the next morning when I woke up, wondering what I was doing in California in the middle of a school year. She didn't buy my vacation line and knew something was not right.

It didn't take long for the mother to find my family's address through directory assistance. I overheard her tell my mother that I was in California and asking what I was doing there. Of course, my mother was shocked and called my dad.

I immediately went on the run and spent the next two weeks crashing at one friend's house after another. At one point, I was sleeping in a pool cabana in someone's backyard. Eventually, I ended up staying at the house of a Skyline High friend, Leddy Zeus (real name). Her father, a colonel in the Air Force, didn't seem too pleased to have me stay there, and I have no idea how they talked him into it. But I had my fake ID and hung out during the day.

It was not too long before the local Oakland Police Department came knocking on the door. As luck would have it, this was the one day when their housekeeper was there. When the police came knocking, the housekeeper came back to my room and asked me my name. Of course, I gave her my alias. But no, the police wanted to talk to me anyway, no matter what name I gave. I think it took all of about ten seconds for the cop to figure out who I was, and he dutifully dragged me down to the Oakland PD, where they tossed me into a holding cell.

The detective, for whatever reason, put me into a freshly painted cell. And of course, what did I do but scratch my name into the clean wall with the zipper of one of my Beatle boots. When the officer came back to check on me, he fumed, screaming at me, "I could have put you in a cell that looks like this one!" as he dragged me around by the scruff of my neck, showing me a holding cell covered with graffiti, scratches, and other assorted marks. Then he dragged me back to my original cell, slamming the steel door so hard I thought it would fly off the hinges. (Note to self: Don't piss off the detective.)

As the hours passed, they would periodically haul me out of my holding cell and ask me if I was going to run again. "Yes," I would reply. I should have lied.

A relative, Sam Meo, came down to visit and tried to relate to me. My family had no idea what had motivated me to take such a drastic step. Although I was being bullied at high school, I had never told anyone about it. I guess it was about a day later when the Oakland PD could not hold me any longer and let me go with Uncle George back to my grandmother's house on Capell Street. All I had with me was a small duffel bag.

After dinner, I said I was tired and immediately went back upstairs, climbed out the window over the garage, and took off again. I guess I was a determined young man.

Two weeks after my escape from Virginia, my father entered combat mode and sprang into action by taking emergency leave from the Pentagon to fly out to the West Coast and drag me back home. The war in Vietnam would have to wait. It didn't take long for him to track me down, and I agreed to a "conference" with him on neutral ground. You would think we were negotiating peace talks. Over the phone, he told me, "Look, I just want to talk to you. If after our discussion you want to stay out there, you can." Later, I found out that the detective handling the case told Dad that he would be crazy to let me stay. Col. Elswick Newport, master Army aviator, was anything but crazy.

We met and talked for a while on neutral ground at a friend's house. While I voiced my grievances about being in Virginia, it was obvious in the end that I was going back with Dad. He firmly grasped my arm, saying, "Let's go home," and that was that. My unauthorized vacation was over.

On the way back to my grandmother's house, where he was staying, he told me a couple of things, one of which didn't make sense at the time. First, he said that my mother had almost died giving birth to me. I had known it was a difficult childbirth because my body was too long for my mother's womb, but I had not known how much physical trauma my mother suffered and that it had nearly killed her. Then he told me that he had had men die in his arms in Korea. This I did not doubt, knowing of his experiences flying an ambulance helicopter for the 4077th MASH during the war.

One thing I always remembered was that he was awarded the Distinguished Flying Cross for one especially harrowing rescue mission:

> ...heroic achievement while participating in aerial flight in the vicinity of Nojon-pyong, Korea. On April 1, 1953, Captain NEWPORT flew an Army helicopter ambulance through a heavy barrage in total darkness to evacuate a seriously wounded soldier from a position close to the front. Guided only by map coordinates and a flashlight, Captain NEWPORT piloted the helicopter at an altitude of only one hundred feet through a narrow valley and landed on an emergency landing pad lit by small quantities of gasoline in cans.... NEWPORT returned to the treatment center across the same hazardous mountain route.

During our trip to the airport, my father complimented me on my escape planning. He was impressed, and no doubt saw a future Army officer in me.

Following our return to Virginia, Dad backed off a little. There were no more forced trips to the barber shop, and I laid low. I went back to school and tried to fit in. It was not easy. There were a couple of minor "excursions" for me when I would simply disappear for two or three days, staying with local friends and hanging out.

On one such foray I went with our roadie (I was playing in a rock band at the time), who had planned with his girlfriend to sneak into her house during the night. I had no place to stay and ended up sleeping in an empty shipping carton in a nearby house that was under construction. I think this was in Annandale, Virginia, in 1968. As I struggled to sleep and stay warm, the next thing I knew I was hearing voices in the empty house. I had picked the one day when a real estate agent was showing the still uncompleted residence to a prospective buyer. All I could do was stay as still as possible and remain undetected. Fortunately, they did not stay long. At that point there was nothing for me to do but wait for our roadie. I had no way to contact him because there were no mobile phones at the time, and I certainly could not bang on the door of his girlfriend's house because it would have tipped off her parents.

I ended up walking around the perimeter of the house, and who did I run in to but the FBI agent who lived next door. His kids were outside playing, and he wondered who this strange, long-haired hippie was in the area. He confronted me and I made up a lame excuse that I was searching for a lost piece of jewelry. He knew I was lying and took my name, then told me to get lost. I walked the several miles back home, and when I arrived my mother notified me that Dad was not going to waste any more time looking for me. OK, so at least we had an understanding of sorts.

By some miracle, I managed to graduate from James Madison High School in 1969. This was after I was suspended multiple times for the length of my hair, forced to attend summer school twice, and stoned on weed during my final math exam. Just going to high school was a process every morning because I had to keep my hair off my ears and collar. I put so much Aquanet hairspray in my hair before school that it didn't feel like hair anymore, but strings of concrete. But I managed to make it through. I was a horrible student and cared only about music and girls.

Our graduation ceremony was held at Constitution Hall in Washington, DC, and both of my parents attended. I cannot remember the exact date, but there was another event at Constitution Hall that I attended—a Donovan concert. This time I was not in a graduation gown, but all duded up in an old khaki Army jacket with flowers sticking out of my sleeves, love beads, desert boots, and red, white, and blue bell-bottom hip-huggers. A *Washington Post* photographer (an old woman to us) grabbed our motley group as we entered, told us where to stand, and took a picture of me and my radical-looking friends. I can only imagine the instructions given to her by her editor: Find the best group of freaks possible. The next thing I knew we were on the front page of the Style section. Mom was furious, and I can only imagine the grief Dad got at the Pentagon since my name was clearly listed in the caption. That went over like a lead balloon. By the time I graduated from Madison, I had run away from home three times, experimented with assorted intoxicants, and opened for the Doors at the Alexandria Roller Rink as a member of a local rock band.

What the reader must understand is that during the period I came of age, some young people in the United States were being subjected to an indoctrination by our country's enemies. Some of the counterculture organizations that sprang up during that period were financially and/or ideologically supported by communists. I distinctly remember reading a newsletter from one such organization showing a group of naked youths running through a field, hand in hand, with the caption "Fuck School!" These forces were active in the US at the time. As a result, some young people didn't know what to believe anymore. Personally, I wrote off all this stuff as communist BS because, being part of an Army family, I knew better.

My father and I did discuss these things, and he recounted his experiences during his second Vietnam tour, where he oversaw procurement for multiple branches of our military and was tasked to assist the local economy in South Vietnam. While reading his résumé years later, I noted how he had assisted the local economy in Vietnam by allocating government contracts to designated companies and organizations and supporting the people.

Unfortunately, that did not keep me from being sucked into this lifestyle to a small extent, mostly by experimentation with intoxicants. During high school, unlike most of my friends, I didn't even drink. But it was not long before I joined the crowd and started experimenting with hashish and cannabis. But at the time, there were many other ways to get stoned, such as purposely taking an overdose of motion sickness pills (Marezine). This I did only once, and to my recollection I took eleven of those pills one night and it was not the best experience, causing multiple hallucinations.

My first real experience with such things scared the hell out of me after I was literally force-fed a dose of LSD. The night in question started when I visited a couple of sisters who lived in our Lake Vale Estates neighborhood. One of them had some acid and was willing to give me a dose, but I had to take it that night. This I did not want to do because I had school the next day. But she was very insistent and before long was on my lap tossing the small white tablet into my mouth. Nothing happened for about thirty minutes, but it was during the walk back home that I began to see nonexistent trees and smoke. We all drove down to a new McDonald's that had opened on Maple Avenue in Vienna, and it was then that I observed animated characters flying through the air singing, "We are the acid people." OK, well, if you're going to take a trip, it should be a good one. But I was starting to get scared. On the drive back to my home, I could not help but notice a rainbow of colors running up and down my legs. It was dark outside, which amplified the experience.

After being dropped off at my house, I stumbled up to the front door, trying to remain calm. The lysergic acid diethylamide was throbbing through my body as I remembered what my father had told me previously, that if he ever thought I was addicted to drugs, he would kill me rather than let me live such a fate.

As my fingers touched the knob, the door exploded open in front of me as my mother grabbed me by the arm and dragged me into the family room where my dad was nursing his usual bourbon and water. My mother said to my dad, "Look at the way he is dressed. It is embarrassing!" I was certain my eyes were highly dilated as I struggled to appear normal. I sat down on the orange Naugahyde couch as dad berated me: "Son, you need to start dressing properly. We can't have you going around in such clothing." I was wearing some green creased pants and a regular shirt.

As dad spoke to me, I observed rainbows of colors spewing from his mouth toward me. I stayed mute and listened to his diatribe. After he was done, I meekly said that I was tired and wanted to go to bed. Fortunately for me, he didn't take a close look at the saucers that had once been my pupils. I managed to get upstairs and lock the door of my bedroom behind me.

During the long night, I laid back on my bed and became disconnected from reality. No longer was I in my bedroom, but floating in space with planets, comets, shooting stars, and wavering colors surrounding me. I remember looking at my hands and they appeared to be those of an old man. I scribbled some nonsense on a scrap of paper, trying to record the experience. I was incredibly paranoid and had auditory as well as visual hallucinations. The whole night I imagined that my parents were trying to open the door. But it never happened.

Somehow, I managed to get up the next morning and go to school. I was not well and was exhausted from the experience. I could not wait to get home and get some real sleep. I never took LSD again. Eventually I concluded that the Beatles were wrong. LSD did not enlighten you, "Lucy in the Sky with Diamonds" was BS, and this substance was dangerous.

It would not be long before I became fascinated with something else: motor racing, in particular Formula One, or Grand Prix racing. Sometime after I graduated from high school, my father was kind enough to help me buy a 1969 Austin Healey Sprite. It was British racing green, and I imagined myself quite the driver as I flung the car around corners on Hunter Mill Road in Vienna. I drove the car so hard that I burned out the clutch in six months by speed shifting, where you hold the throttle wide open while shifting. Consequently, I had to learn to work on the car myself and taught myself how to pull the engine and replace the clutch. While I still drove fast, I tried not to abuse the car anymore. I was just a young kid having fun with the machinery.

Because I had a sports car, I read every monthly issue of *Road and Track* magazine and imagined myself a real racing driver. One European driver stood out to me: later three-time world champion Jackie Stewart. He was Scottish and one of the few guys to sport long hair. That drew me to him. In addition, he was very quick and had been a close friend of Jim Clark, who tragically lost his life in a Formula Two car at Hockenheim, Germany, in 1968. I remember seeing the picture of Clark's crumpled Lotus on the front page of the *Washington Post*'s sports section—it was captioned "Death Machine." The *Post*, in particular Shirley Povich, was not a fan of motorsports and said they should be banned.

At that time, Formula One races were not broadcast and only mentioned in passing in the newspapers. The only way to keep up with the sport was by reading the magazines, which I did regularly. One night while thumbing through the back pages of *Road and Track*, I saw an advertisement: "Train and Race in England with Jim Russell." Wow! I could go to England and learn how to race open-wheel cars? Apparently so.

I had few things going for me then. I had worked at mostly dead-end jobs, such as landscaping, on the loading dock for the Hecht Co. at the Tysons Corner Center, running a Gardner-Denver rock drill for Diamond Drill Blasting, then as a janitor. After dropping out of the local community college, I wasn't exactly headed in the right direction. Since Dad had pretty much given up on me and stayed off my back, our relationship had improved. He still didn't approve of my hair length but let it go for the most part. I remained the black sheep of the family, while my brother studied at the University of Maryland and joined the ROTC.

I approached Dad about my idea of going to England and learning how to become a racing driver. To my surprise, he was supportive. While I had traveled extensively with the family, I had never really been anywhere, especially overseas, by myself. Given all the issues we had between us, I can only think he felt it would be a way for me to grow up and become a man.

My parents agreed to a deal: If I could come up with the money to pay for the Jim Russell School, they would cover my airfare. At the time, I was working as a janitor cleaning offices at the Tysons Corner Center in Virginia. This was not exactly a career. I can't remember whether I quit my job or took leave, but I found the time to go to England.

I sent my form and deposit to the school via Air Mail (no electronic mail in 1970) and eventually received a response. I still have my confirmation letter from one Mr. Colin Campbell, dated 18th May 1970, mailed to Curtis W. Newport, Esq.

Dear Mr. Newport,

Thank you for the Reservation Form and also the deposit (cheque $60.00). We note that you will be arriving on 14th July and I have reserved accommodation for you at the Griffin Hotel, Attleborough, Norfolk, which is approximately 3½ miles from Snetterton Circuit.

When you arrive in London you should take a train from Liverpool Street Railway Station to Norwich and then a further train from there to Attleborough. We suggest you telephone this office when you reach your hotel so that we may discuss your training programme. Incidentally, this will commence at 9am on Friday, 17th July.

We hope you have a pleasant trip to England and look forward to meeting you in July.

Yours Sincerely,

Colin Campbell

I was on my way to England and fame and fortune!

4

1970: England

My parents arranged a low-budget charter flight to get me to London, where I landed at Gatwick Airport. At the time, Gatwick was new, having just been completed. I carried my US Army Air Corps B-4 Valpak and a small handbag. Somehow, I managed to get on the right train to London, and before long I was wandering around the streets in the typically dismal English weather trying to make my way to Victoria Station.

I must have looked lost because before long, an old geezer who looked like something out of a Charles Dickens novel grabbed my bag and deftly hoisted it onto his shoulder as he asked, "Where to, Guvnor?" I was taken aback. Was he trying to rob me? What was going on? I decided that he was just an old bum looking to make some easy money off a Yank. The two of us chatted as he carried my bag to the station, and when we arrived he set the bag down, obviously looking for money. In 1970, the money system in England was complicated to an American, and I think I gave him a few shillings. He tipped his hat to me and went on his way.

After arriving in Attleborough, I stayed at the Griffin Hotel, a few miles northeast of the racetrack. The Griffin was typical of small country hotels of the day: rooms on the second floor as small as closets, a loo down the hallway, uneven floors with threadbare carpet, and old plastered walls that looked anything but flat and straight. It was originally a coaching inn, built in 1560 during the reign of Queen Elizabeth I, and had wooden cobblestones around the front so visitors could arrive on their horses and not disturb the guests. Most of the furnishings dated from the 1940s or 1950s, and it had a basement with large iron rings still attached to the walls from when soldiers would chain up their prisoners while en route to the courts of assize. There was rumored to be a tunnel running from the basement to a doctor's house in town, but the reason for its existence was unknown. The nearby parish church dated from 1358.

I was tired and hungry from the trip, and before long I was enjoying a typical British meal of some sort of meat and potatoes. I spent the next day or so exploring the town and hanging out at the pub down the road from the hotel. I knew that the Jim Russell School was sending a car for me and other students early Friday morning, so I just tried to relax and get used to the time change. The Griffin had a common room downstairs with an old black-and-white television set where the guests watched BBC 1 and 2 together every evening. It was quaint. I made a collect call home using a pay phone and told Mom and Dad that I had arrived safely and not to worry.

Precisely at 9:00 a.m. on Friday, I and several other hopefuls were at the track getting our first glimpse of what looked like an orange plastic missile with wheels. It was so small. But in that we were supposed to learn how to race cars, but not necessarily become racing drivers because that accomplishment involves far more than simply knowing how to go fast. Being a "driver" requires so much more: strategy, lack of fear, dedication, and maybe a willingness to risk your life.

Before me sat an impressive-looking Lotus 51C Formula Ford, identified as number 3, with treaded radial tires, a 1600 cc engine, Hewland gearbox, a few gauges, a gear shift, and radiator. Anthony Bruce Colin Chapman's emblem was prominent on the battered fiberglass nose, which looked as though it had been used for target practice with a shotgun.

Missing were the seat belts. The Jim Russell International Racing Driver's School found them to be an encumbrance because with them it took too long to get in and out of the cars during training. No one mentioned what would happen if you crashed. Such incidents were described by a variety of British slang terms, such as "shunt" or "prang." They all meant the same thing. You hit something that you were not supposed to. There were rumors about other driving schools, about how students had crashed and been injured or died, but these were urban legends.

Looking back at it now more than fifty years later, I realize that my route to standing with a bunch of strangers on the tarmac at Snetterton Circuit in England on July 17, 1970, was a torturous one. Growing up, I thought that anyone who would race cars was insane. Risk your life for what? A trophy? Money? Fame? Women? I could not see how it would ever be worth it. But when I got my 1969 British racing green Austin Healey Sprite, then I knew. Then I understood. Going fast could be fun. I already had a Restricted Racing License, issued by the Royal Automobile Club, London, in care of Jim Russell Racing Driver's Club, Snetterton Circuit, Norwich.

While I flung my version of a Grand Prix car (an Austin Healey Sprite was my foil of speed) around the back roads near Vienna, Virginia, just after high school, I sported a tweed cap, a mustache (to look like racing driver Graham Hill), some sort of jacket tied to motorsports, and leather gloves. There was not a Vienna cop that could catch me in their Ford Crown Victorias. I fantasized about winning the Monaco Grand Prix and standing at a podium while our national anthem played. I watched John Frankenheimer's movie *Grand Prix* endlessly. I collected a bunch of new heroes, such as Jackie Stewart, Graham Hill, Jacky Ickx, Jochen Rindt, Piers Courage, and Jo Siffert. Many of them would later die on the track.

Twenty-five years after the Eighth Air Force's last B-17 bomber took off from Snetterton to rain destruction on Germany, I was standing there shivering on a cold morning, nineteen years old, among a few British guys, a Swede, a German, and an older American, because for some reason I, too, wanted to race. The Swede was thin, with blond hair, reportedly an engineer with a wife and kids back home. The German was older and looked like he belonged at the controls of a Messerschmitt Me-109; he was a writer doing a piece on motor racing. Their names have faded into memory, but a few can still be resurrected from the grayness that clouds all things that happened over fifty years ago.

Phil Heim, in his gold Nomex driving suit, was probably in his mid- or late thirties, a product of what Hemingway once said: "There are only three sports: bullfighting, motor racing, and mountaineering; all the rest are merely games."

He had already spent time in Spain trying to become a bullfighter, though it was unclear as to whether he actually did it. Now he was trying his hand at racing. Phil was a rugged-looking individual with a face that would have looked good on camera. He seemed to have money, because he wore a massive Rolex and was able to pursue such things without the pressing need of a job. None of us were sure what he actually did, though he once shared with us that he had been in Vietnam during the war as a "businessman." I can only assume that he was too old for the draft or had found a way out of it.

Another American was Leigh Gaydos, who was about my age and apparently had already completed his training at the Jim Russell School. Leigh was painfully thin, had a promising mustache, was very quiet, and wore a tweed racing cap similar to mine. He was from Binghamton, New York, and the result of a middle-class upbringing. But he was serious about racing, and it was all he thought about. I remember he had an older brother who was working in England as a professional musician. Leigh was hanging out, hoping for his big break.

Other than that, I knew little else.

The lot of us purchased stiff new driving coveralls, accordingly flame retardant, which we excitedly pulled out of plastic bags and donned with difficulty in the school's paneled lobby. Mine was several sizes too large and folded where my joints bent. We accepted our battered blue-and-white crash helmets and goggles, eager to get to the track to prove what we could do at speed.

John Paine, our instructor, had us all on the pit straight as he described how to take Riches Corner. "You want to leave your braking until you hit this point," he told us, with his hand motioning down toward a spot where, if we had enough bravado, we would be going flat out, until we hit that mark. John had flushed cheeks, was of medium build, and wore a pale blue driving suit and driving jacket—another man who had not made the cut as a professional driver and ended up a teacher. The cars, all of them multicolored Lotuses, were strung out in a staggered line along the track, waiting in anticipation for our inexpert handling.

"After you finish braking, you want to keep your speed steady, neither accelerating nor decelerating, for maximum adhesion," we were told.

I raised my hand. "Doesn't a car have the best grip when it's accelerating?"

"No, keep the speed steady through the corners. Off you go . . ."

We awkwardly climbed into our low-slung tubular vehicles, started up our Ford Cortina engines with a puff of black smoke, and began driving around the track, trying to remember our instructions.

That first experience of driving such a car was like having your entire existence rocket through a funnel of speed. It was not like piloting any sort of regular automobile, but being propelled along the tarmac in a large go-kart. Every imperfection in the pavement was instantly absorbed by your body, hands, and arms through the tiny vibrating steering wheel. You heard and felt the rushing of the wind against your face, the wail of the unsynchronized gearbox, and the sound of the rolling tires against the track. The whole car shook and bounced while you tried your best to keep it on line and on course, as we all tried to hit our apexes, the crown jewel of any corner. Shifting gears was akin to staging in a Moon rocket. You grabbed the spherical end of the shift lever, racked it out of third, and slammed it into fourth, while your body felt every bit it of it, twisting around in your low seat, left foot pressed hard against the footrest, in a desperate attempt to stay in place and control the beast you were driving. The wind and engine howled in the background, a constant reminder as to how fast you were really going. The tall earthen berms on either side of you blurred, but you kept your eyes straight ahead, trying to hit that braking point and corner just right, so John would think you had it in you. The Lotus stopped and started faster than anything I had ever seen (and I thought I was pretty hot in my 1969 Austin Healey Sprite on the Northern Virginia country roads). You smelled everything—the petrol, the hot oil, the dripping sweat on your face, and the fear. You felt alive.

Round and round we went over the days, and I'm not sure anyone was even recording our lap times. At that point, it was more about getting the feel of the car, learning the course, and slowly working up to a reasonable speed.

In the evenings, after a hard day of learning, we'd all meet at a pub across the street to share stories and drink copious amounts of warm beer. The locals accepted us with some amusement I think because they'd seen it all before: young lads, with delusions of glory, hanging out at their pubs while we went through our training at Jim Russell. We drank, laughed, and shared our stories of the day. The females were a bit wary of us strangers, and the most they were willing to offer was some conversation, acceptance of a free drink, and maybe a small kiss if they liked you. I can only imagine that it was much the same way when painfully young American airmen invaded their existence during the war. We were both an invasion of sorts, spending money, and looking for action in a small village where everything shut down so early.

Part of it was simply learning the British customs and terms for things. And the confusing monetary system didn't help. Cars were described as having "bonnets," "boots," "gearboxes," and "wings." The money was pennies (massive affairs, most of them minted in the nineteenth century), tuppence (twopence), thruppence (threepence), shillings (made up of twelve pence), and quids (a British pound). Eventually I learned it all. They didn't have wake-up calls then but "knocked you up" in the morning by banging on your door. Instead of refrigerators in our rooms, we stacked our milk, cheese, and butter on the window ledge.

Every day, a small van from the school arrived to haul us to the track so we could start driving. The training continued throughout the week under John's watchful eyes. Sometimes, I'd catch a glimpse of him lurking at one corner or another, keeping an eye on us, taking notes, then bouncing off in his Vauxhall school van. It didn't matter whether it was cold, dry, wet, or a downpour—we drove. Because in England, if you wait for good weather, you could be waiting for an eternity.

We'd drone past the pits, the cinder block cubicles now inhabited by aging motorcycle racers with their scruffy beards, thick accents, and worn-out black leather racing garb. Some of them looked like they were barely capable of driving a car, much less a British racing bike. But we all existed out there together, them holding us up in the corners, then blasting past us in the straights.

It became a bit monotonous: passing the pits and under the bridge into Riches Corner; grinding the gears while downshifting to third and trying for that smooth line; a short breather before Sear, then upshifting to fourth, then fifth, a gentle bend to the left, then opening it up for the Norwich straight; on the brakes hard entering the hairpin, down two gears into third, trying to keep it all pointed in the right direction while methodically going back up to fourth; a quick jog through the Esses, a smooth line through Coram, quick blast of power, and threading the needle of Russell Corner before the blur of the pits and grandstand; then doing it all over again, round and round.

I was still driving too slow. As soon as I hit the Norwich straight, I didn't have it in me to keep the throttle fully open. To do so was like unleashing a wild beast that I had little confidence in taming. I was scared—scared of how fast the Lotus bullet was going, scared of my young ability to control it—and starting to wonder why I was doing any of it. As I gained speed in the straights, I looked at my hands on the vibrating steering wheel, felt the bumpy tarmac, and was not convinced I had the courage, guts, or ability to drive at 10/10ths—in other words, full out. It was inevitable that I would have to make that judgment, cross that line, and see if I could do it. I was like the guy lying with his face in the French sand on Omaha Beach on D-Day. Did I have the mental strength to stand up, to take the risk? There was only one way to find out.

How many laps I completed before the inevitable crash have been lost to the ages. I held my foot down hard on the throttle pedal down the Norwich straight, maybe braking a bit later, extending myself through the corners, and trying to keep things under control. There comes a point when you've gone just a bit too far by asking just a little too much of the grip of the tires, as well as your skill in managing it all. That happened to me at the exit from Russell Corner, just before the pit straight. I thought it had been going well, but after negotiating the apex and while exiting into the straight, I watched in horror as the steering wheel went limp in my hands. I was trying to turn left, but through the physics of mass and tire friction, that car was going straight, right toward an earthen wall. Granted, the car was not outfitted with racing tires, but the same street radials anyone could buy at their petrol station. Had I known more, I would have worked the tires back and forth in understeer and floated on that edge of adhesion. But I didn't.

With the wheels turned left and the car going straight, it was much like a famous scene from *Grand Prix*, where actor James Garner leaves the Monaco track in his BRM Grand Prix car and ends up in the Mediterranean Sea. But this wasn't a movie. It was real. And I was not James Garner.

I smacked the side of the dirt and grass wall and remember the right front wheel ripping away from the suspension, sailing somewhere past my head and into the air. As the car careened along the top of the berm, I was basically just hanging on, trying to keep myself from being ejected as my body was tossed back and forth within the confines of the steel tubular frame. It was like surfing a wave of exploding dirt and grass. Eventually, I stopped and had the sense to turn off the hot engine. Silence. The still warm exhaust pipe crinkled as it cooled. I crawled out from the mangled mass of fiberglass, rubber, and metal, then saw John bolting down the straight from the pits as I hobbled away from the scene with my helmet and goggles in hand. My right leg throbbed from banging against the frame of the car.

"Are you OK?" he asked.

"Yes, I think so, John. It was the steering."

I honestly believed that statement. I could not see any other way the car could have done what it did, but then, what did I know? As I limped back into the pits, I watched John check out the car then angrily head toward me, his face stern and flushed.

The veins in John's neck were bulging as he yelled, "There's nothing wrong with the steering. You were going too bloody fast!" With that he stormed off to the office, leaving me staggered, saddled with my guilt.

I sat in the pits with one of the geriatric motorcycle racers.

"John's a good bloke," he said. "He's just mad that you blamed the car."

I sat with my head in hands, my self-confidence, bravado, and honor deserting me, in search of a more worthy owner.

There followed a driver's meeting at the office, with all of us sitting in a semicircle on the benches in the dressing area. Then Jim Russell himself came in: three-time winner of the British Formula Three Championship with seventy-five victories, a technical advisor to the John Frankenheimer film *Grand Prix*, and sponsor of Brazilian Emerson Fittipaldi, twice the Formula One world champion.

Jim was now a stout man pushing fifty, hand-rolling a cigarette, with arms disfigured by scar tissue. (A chance encounter with flame?) He began.

"Right, so we've had a prang. Who was it?"

I sheepishly raised my hand, forcing it to be steady.

"It's not about going fast right now. It is about learning the track, gaining confidence, and becoming comfortable in the car. You'll get quick enough later. It's going to cost us about 500 pounds to repair that car, so let's not have any more shunts, OK?" He was looking right at me.

I did the mental calculation and figured that between the car repair cost and my school fee, the Jim Russell International Racing Driver's School might break even in the end, with me as its expensive student. With that, we went back to driving.

I was back in the car, circuiting the track. But whatever I had, or thought I had, was gone. I was simply going through the motions, driving the lines, shifting the gears, no longer with anything to prove. There was a second, though minor, encounter with John Paine, as I took a sweeping line through Riches Corner. It had been raining and the track was wet. As I hit the apex, the car wiggled—not a lot, just a little. I kept it straight.

John confronted me later. "You almost lost it, didn't you?"

"No . . . it was fine. I was being very careful. The pavement was wet."

John was always going to be watching me from now on because I had failed.

The climax of the school was a student race. By then, my confidence had shaken. I finished dead last. But I didn't crash another car.

With training completed, our class hung around Attleborough, the track, and the pubs, trying to figure out what to do. It was too early for me to go home since I already had a scheduled flight. So, a few of us decided to embark on a tour of the most famous racetracks in England and continental Europe. Plus, we knew that the Austrian Grand Prix was going to happen and possibly we could make it there in time.

It was just before our trip that we ran into Rikky von Opel. Rikky was a flamboyant figure who owned two Formula Fords: one a Lotus 61 and the other a Merlyn, sometimes referred to as the "flexible flyer" due to the frame's lack of rigidity. None of us seemed to know what he did for a living, except that he had mentioned he was an art appraiser. While Rikky was born in New York City, he was the son of Fritz von Opel of automotive fame, an early experimenter of rocket-powered cars in Germany in the late 1930s. Of course none of us made the connection of his last name to the Opel car company at the time. He was just a funny guy who wore a red bandanna and seemed to not have any problems spending money. Apparently, his dad thought he was going to school in England, not pissing away his allowance on racing cars. Consequently, he raced under the pseudonym "Antonio Bronco." He didn't like having his picture taken for that reason, though I have a few.

We ended up being his pit crew by helping him get the car on the track, strapping him in, and giving him signals while he raced. Rikky was quite a good Formula Ford pilot, especially in the rain. I cannot remember how many races we attended, but he either won them or finished in the top three. Formula Ford in England was then, as it still is, very competitive. There weren't any slackers finishing at the front of the pack. You had to be good to be there. And Rikky was good. From what I know, he eventually ran Formula Three, winning the Lombard North Central Championship in a Lotus (61 points) and finishing second in the Forward Trust Championship, a close second behind England's Dave Walker (another hanger-on at the Jim Russell School). Later, he bought himself a ride in an Ensign Formula One car and drove for Brabham as well, eventually competing in fourteen Grand Prix races. But with all his money and skill, Rikky never went anywhere in Formula One racing. In fact, at one point in time his fellow drivers tried to have him banned. As one racer stated:

> Accusations started floating around that von Opel was no more than a play-boy racer who had more money than ability. Vagner Salus said that he heard that some of the other drivers petitioned to have Rikky banned, although there was no evidence to support this. In all fairness, von Opel was not that bad at all, and he took his F1 project very seriously. The Liechtensteiner [von Opel] answered his critics in the best possible way at Zandvoort, recording the 14th fastest time in qualifying, only 2.54 seconds behind Ronnie Peterson's pole-sitting Lotus.

Still, Rikky von Opel was a very talented driver from what I saw. What's interesting is that with all his perceived skill, he barely qualified in Grand Prix races. It makes you think about how good the drivers and cars are that place and win . . .

After returning to the States, Leigh Gaydos and I stayed in touch via regular mail. He was going to the US Grand Prix and suggested that I come up and see the race with him. Leigh also mentioned his friend Spanky Smith, who had a Winnebago, was a well-known race photographer, and might let us sleep in his van during the race.

It sounded fabulous. I took a Greyhound bus to New York and Leigh picked me up from the terminal. He was driving an old piece of Detroit iron, a sedan of sorts, with a big engine. I think it was the family car. Upon arriving at his home, I could tell that he came from a typical middle-class family.

His parents' house was a simple affair, a small single-family home, cluttered with articles and small bedrooms. They were so short on space that Leigh and I had to sleep in the same bed, head to foot, as was done in the day with men.

The next morning, we made our way up to Watkins Glen, with Leigh flinging his family car around the numerous bends with abandon. The trip was filled with screeching tires, a roaring V-8 engine, and the smell of hot brake pads. Not surprisingly, along the way we got pulled over by a state trooper on his way to work and Leigh got a speeding ticket.

I will never know the particulars of how Leigh and Spanky originally met, or even the depth of their relationship because both men have been dead for decades. However, it seems likely that Leigh, as a teenager, was hanging around Watkins Glen professing an interest in becoming a racing driver. Spanky was always at the Glen because he lived not far away, with his mother, while working at IBM. His mother called him "Edwin" during the one time I stayed with him in Binghamton, myself resigned to the famous Winnebago parked in the backyard.

Leigh introduced me to Spanky during the US Grand Prix in 1970, not long after I returned from the Jim Russell School in England. Looking back on it, I think Spanky was a bit wary of me during our first meeting because at that time, I could be painfully quiet and shy or an arrogant prick who thought he knew everything. I was usually somewhere in between. I knew little about Spanky, other than he photographed racing cars, knew all the drivers, and had a Winnebago that was outfitted like a racing museum, with assorted images of drivers, cars, and racing posters attached to every square inch of the interior. One thing was certain: He was well plugged in to the international scene and was especially tight with the McLaren Team, in particular Bruce McLaren and Denny Hulme, both from New Zealand. Through him I eventually met Hulme, Peter Revson (of Revlon fame), Jackie Stewart, and others. As one racer later recalled:

> I don't even remember Spanky's last name. I don't think anybody ever knew it. There would be Denny [Hulme], George [Follmer], Revvie [Pete Revson], and the photographers, and Lothar and me, and the wives. Jackie Oliver would be there. We'd always gather at Spanky's trailer, and everybody would sit there and laugh and shop talk. They'd say, "You should have seen me in this turn!" and just bench race.

Spanky was an affable individual, sporting substantial muttonchops and a sizable waistline. It had to be a struggle for him to lug two heavy 35 mm cameras, a bag, and film all over the Glen, but he did it for decades. After the race, one of the first things he did was mail me a handwritten questionnaire of sorts, which apparently was something he did in an attempt to get to know people. While the contents are mostly lost to memory, I do recall that some of the questions were quite personal and hilarious.

As Spanky and I corresponded by mail, I learned more about him. At the time, my father was serving in the Vietnam War, and along the way, I learned that Spanky had served in the Army during the Second World War in Europe, in particular the Hürtgen Forest during the Battle of the Bulge, the longest battle on German ground during WWII and the longest single battle the US Army has ever fought, resulting in 33,000 US casualties. He also revealed to me that during the engagement, he was shot in the neck by a German sniper and spent many months recuperating.

Leigh and I stayed in touch by mail following the US Grand Prix at Watkins Glen. Apparently, he spent a lot of time driving a UPS delivery van to make enough money to buy himself a ride in a Formula Ford fielded by the Jim Russell School. There was also talk about him getting sponsorship from Levi's, the clothing manufacturer. Leigh was serious about racing in a way that I never could have been. I think I was only playing the part, while Leigh meant business. In one of the letters he sent me from England in March of 1971, he discussed my crash at Snetterton and how I should feel about it:

> I think I know the feelings you are speaking about after crashing. Obviously, when a crash occurs, if it was not a mechanical breakage, then it was most likely driver error. . . . You must admit to yourself that you have made an error, but then you must be back in the car and try not to make that error again. . . . You must try not to let it blow your confidence completely, you must just try and work through such things. . . . You can do a lot in your Sprite by race driving on public roads. . . . You know that already; keep at it until you're in a proper racing car again.

When I read this letter today, it crinkles like an aged document of old, having the feeling of parchment. A lot has happened in the fifty-plus years since he wrote it. The letter was typed, so Leigh must have purchased a typewriter in England to send requests for sponsorship to various parties. Only serious men used typewriters in those days, which tells me Leigh was consumed by the idea of being a professional racing driver.

Leigh eventually gave Jim Russell 1,500 British pounds to buy himself that ride in a 1971 Merlyn MK20 Formula Ford, with a "full-race Holbay engine" with the "latest Firestone Torino tyres." It was the last money he ever spent on racing.

Three months later, I sat in the family room of our Vale, Virginia, house reading the sports section of the *Washington Post*. I normally scanned it for any motorsports news. As my eyes crept down the page, I noted the following obscure article:

LEICESTER, England – (UPI) – American racing driver Leigh Gaydos was killed yesterday when his car crashed during a Formula Ford race at Mallory Park track in Leicestershire. Gaydos, a 20-year-old bachelor from New York City, crashed into a marshal's post on the 13th lap of the 15-lap race. Ambulance men rushed to the scene, but Gaydos was dead when they pulled him from the wreckage of the Jim Russell School Merlyn he drove.

I recall turning pale at the news. My blood pressure spiked, my heart raced, and I felt faint as I tried to comprehend what I had just read. My mother asked me what was wrong. I didn't bother answering. I left my house, climbed into my Austin Healey, and simply went driving. I don't know where I went. I just needed to be out somewhere. To be out driving.

In 1971, communicating with England was not a trivial task. There was no email, and overseas telephone calls were a challenge at best. But I wrote to Colin Campbell to try to find out what happened. His reply:

What actually caused the accident at Mallory Park I just cannot say. From what I can gather, he pulled out to overtake a car immediately in front and apparently lost control, hit the grass verge, came back onto the track, and crashed sideways into a Marshals Post. The coroner's enquiry resulted in a verdict of "accidental death." . . . Leigh was particularly popular with everyone at Snetterton and Attleborough: he was always so pleasant and friendly.

Spanky Smith was more forthcoming. He told me that Leigh's car had hit the concrete marshal's post so hard the front wheels were touching the rear. His demise was sealed. This shook me in a way that I had not thought possible. It was like having your best friend's head taken off by a bullet in the war: If it could happen to him, it could happen to me.

Before Leigh's death, Jochen Rindt had already been killed while practicing for the Italian Grand Prix at Monza. He bled out on the track when his jugular vein was cut after an inboard brake shaft of his Lotus 72 car snapped, flinging him into the guardrail. He was only twenty-eight years old. Jo Siffert, the shy, friendly man who gave me his autograph at the Österreichring, Austria, while staying at a local bed and breakfast, was incinerated in his BRM Formula One car at Brands Hatch in England the same year Leigh died; the magnesium frame of his car burned like a sparkler. A couple of years later, François Cevert had his body literally cut in half after hitting the guardrail at the same track where I had enjoyed the Grand Prix with Leigh and Spanky—Watkins Glen, New York. These were men taken by either mechanical malfunctions, their own errors in judgment, or simple fate.

I have concluded that there are men who are compelled to prove themselves in some sort of conflict, whether it be through sport or combat. They are willing to risk everything, whether it be dodging bombs or pushing their car to the limits in the rain. What more could you risk than your life? What more could you gamble with? How else could you possibly prove your skill than by balancing your life against your abilities? My friend Leigh, as well as Rindt, Siffert, and Cevert, played the cards they held and lost. Even so, after all of it, I still wanted to emulate these heroes of mine. I just didn't want to die in the process.

I continued in my quest to become a professional racing driver by purchasing an aging Beach Formula Vee racing car to compete in local races sponsored by the Sports Car Club of America (SCCA). The car was pathetic: dry-rotted Firestone racing tires, a laughable cable system in the rear to control the camber of the wheels, a fiberglass fuel tank that doubled as a seat (and always leaked gasoline), and a blunt fiberglass nose that had been cut off because the car was too long according to the rules.

At the time I was working at a local Volkswagen repair shop called The Bugworks. At least it gave me a chance to learn about VW engines and collect the spare parts I would inevitably need to keep the car running. I managed to get through two SCCA driving schools and competed in several regional events at the Summit Point Raceway in West Virginia. By then, my father had returned from his second tour in Vietnam and left my mother. All I knew was that he returned home, packed up his things, and left while Mom was in California visiting her sisters. Dad tried to talk to me about his decision, but I was angry with him for breaking up our family and spent my time working on the Beach car. I wish I had talked with him because it probably would have helped me to better understand his feelings.

I recalled a comment he had made while he was dragging me back home following my runaway: "It was a marriage that never should have happened." I didn't understand what he meant at the time, but decades later, I did. After his death and while going through his papers, I concluded that my mother was several months pregnant with my brother when they were married in Reno, Nevada, in 1948. In other words, he had to marry Mom because in those days, it was something men did. There was no way in hell Dad could get out of marrying our mother. Not with her brothers. George, Charlie, and Johnny would have taken him to task, and he knew it.

Surprisingly, Dad was supportive of my racing efforts before he left. He accompanied me to Summit Point and shot some 8 mm film of me struggling with the Beach Formula Vee. I continued to drive the car for a couple of seasons, extending my experience to Virginia International Raceway (VIR) in Danville. However, I was still having problems with the fuel tank, and during one race, I felt gasoline leaking down my back and pooling in the bottom of the seat. As soon as the race was over, I pulled into the pits and shoved a garden hose down my back to dilute the car's fuel. For many months after that incident I had burn marks on my back.

Eventually, I decided that if I wanted to be competitive in Vee racing, I needed a better car. So I sold the Beach and purchased a Zink Formula Vee from a local driver. The difference was striking, and I immediately started logging some decent lap times. I was told about the car by a local engine builder and racer in his own right, Bob Houston. He and a local real estate developer, Kip Laughlin, were making a serious attempt to win the SCCA National Championship in their car, and they were kind enough to take me under their wing. It helped that Bob had originally built the Zink from a kit. It was also raced by other prominent drivers such as Bill Scott and Glenn Sullivan, who drove it at some of the famous Vee races at Daytona Motor Speedway. So, I knew it was a good car.

I drove the Zink in many regional and national races. At Summit Point, I managed a second-place finish in a national contest while running a half a second under the existing lap record. It was a hard-fought battle, with the third-place car all over me the whole time and us trading spots in the draft down the main straight. Summit was a good track for a Formula Vee, especially if you had the courage to keep your foot full down on the throttle down the chute, a very fast downhill right-hand corner terminating in a slow left corner. I admit, it was scary.

One competition stands out to me, which was during a regional race at VIR. I later set a new lap record there for my class that stood for many years, mostly because the track closed for decades before reopening. At the time, VIR was a dangerous track because it was very narrow and bumpy, with numerous areas where if you went off, you were headed into the trees.

I was duking it out with a United Airlines flight dispatcher named Don Taylor, also driving a Zink. Going into the hairpin corner before a long straight (almost a mile), I was on his tail, pulling up alongside of him as we accelerated down the back straight. As I drew alongside, he started crowding me to the left side of the track, not giving me racing room. This angered me. As our two cars sped down the long straight section, I decided to make him move over by placing my right front wheel in between his left front and rear wheels. This was a dangerous thing to do, especially at high speed, because if our wheels touched, one of us was going to get launched into the air. Don didn't move over, and I got angrier. We were going even faster, easily over 100 mph, headed toward the downhill esses before the pit straight.

For whatever reason (maybe my stupidity and youth), I began zigzagging, trying to force Don to the right, shoving my wheel in and out of his. My maneuvers became so extreme that I somehow veered off the left side of the track and the next thing I knew, I was plowing through a cornfield at full speed. All I could see were blurred visions of corn stalks whizzing past my tiny race car at immense speed as they flew into the air. I was riding on a sea of corn, just barely in control. Somehow, I managed to keep the Zink pointed straight and exploded back onto the track in a tsunami of corn, leaves, branches, and squealing tires. Don was by that time well out in front of me because I had lost speed during my off-course excursion.

I really got the bit between my teeth as Don probably figured I was out of contention. I could see him ahead of me. Maybe, just maybe, I could catch him . . .

Indeed, I caught up to him in the downhill esses, drew abreast of him at the finish line, and beat him by 2 feet. The entire front suspension of my car was dragging corn stalks the whole way. I don't even remember what position we were fighting over, but it didn't matter—it was a position.

Many decades later, I raised the subject with Don during a party at Bob Houston's house. Incredibly, he didn't even remember the incident. I could have easily wiped both of us out and remembered it like it happened yesterday. But I do recall him saying after the race that he was "surprised that I wanted it that bad."

To finance my racing, I was still working for Diamond Drill Blasting out of Manassas, Virginia, running a rock drill. I didn't really like the job because it involved being peppered in the face all day with quartz dust and having oil drip on my head from the hydraulic drill motor above. But it was a good-paying job at the time: $3.25 per hour.

Once, we were blasting for an extension of the Bethesda Naval Hospital. While it was a union job, none of us were union members, which did not endear any of us to the other workers. In fact, our supervisor instructed us to run our drills at half-speed to milk the job for everything we could. My first day on the job gave me a splitting nitroglycerin headache because I neglected to wear gloves to keep the chemical from being absorbed into my skin.

Even though I was not a licensed blaster, or "powder monkey," I still set up all the charges for whatever we were doing. This involved priming the dynamite, usually 2-by-16-inch charges with electric blasting caps. (Sometimes we strung the sticks on long lengths of primacord, which has a core of PETN [pentaerythritol tetranitrate], a very fast-acting explosive.) I also had the difficult task of stemming the holes with large boxes of gravel. It was exhausting work. One day during our work on-site, I discovered that I had primed one charge too many. In other words, there was a stick of dynamite sitting on the ground outside of the hole with an electric blasting cap inserted. This was not good due to the possibility of the cap being set off by radio signals or static electricity. Consequently, I decided to "safe" the charge by removing the blasting cap.

These charges were primed by sticking a small rod in the end of the dynamite and inserting a blasting cap, which was secured by a half hitch using the wires to the cap. I was having a difficult time removing the cap and recall that at one point, I had the stick of dynamite under my armpit while trying to yank out the blasting cap by the wires. After about a minute of struggling with the stick, I had a light-bulb moment as I thought to myself, *Maybe this is not such a smart thing to do* . . . I put the stick back on the ground and let someone else deal with it.

I think the blasting company sensed my lack of enthusiasm for the work and eventually laid me off. I was unemployed again.

Several years earlier, my father had given me a very special booklet from the Pentagon that would change my life. It was titled *Undersea Vehicles for Oceanography*. I devoured the text and diagrams, reading about several manned submersible systems, such as the Aluminaut, Deep Quest, and deep ocean work boat. The booklet detailed the current state of the art in underwater systems. I distinctly remember watching *Sea Hunt* on television throughout my grade-school years and wanting to go underwater myself. But I did not sign up for scuba diving lessons; instead, when I was thirteen, I started designing my own diving gear.

My first creation was simple: a plastic bag fastened to the end of a section of garden hose fitted with a snorkel mouthpiece. At first I didn't know anything about oxygen partial pressures or carbon dioxide buildup and consequently rebreathed the air initially contained within the bag, which caused me to get dizzy underwater a few times. I later modified the rebreather with a flapper valve (made from a rubber party noisemaker) and a larger bag so I could exhale my own breath.

My second attempt at building dive gear used a standard fire extinguisher (the type that is charged up with low-pressure air). I modified the valve on the tank so that the air contained within the tank would flow out at a constant rate and manufactured a breathing bag and tank harness out of an air mattress and old parachute webbing brought back by my father from the Korean War. Fortunately, I only used the tank in a shallow swimming pool and didn't manage to kill myself. Knowing what I know now about pressure-related physiology, I am lucky to have survived my experiments.

During that time I was also designing submersible vehicles. They were elaborate designs using aircraft wing tanks for pressure hulls and electric motors and automobile batteries for propulsion. I remember once writing to Perry Submarine Builders (now Perry Technologies) in Riviera Beach, Florida, asking for information about their manned submersibles (I think I was nine or ten and must have scrawled out some kind of letter). I received a reply from John Perry that included several brochures on the company's diver lockout submersibles.

Past submarine builders also held a fascination for me. I read several books about John Holland, the builder of the first submarine ever commissioned by the US Navy, and Simon Lake, another early submarine builder who later sold submersibles to Russia. The whole concept of traveling and working underwater was what intrigued me. It was not only being underwater I was interested in, but the hardware required to get you there—the ballast tanks, hatches, viewports, pressure hulls, dive planes, and propulsion motors.

I finally ended up in the commercial diving business in 1974 by answering a help wanted ad for diving systems technicians in the *Washington Post*. The company was Ocean Systems Inc. and I was twenty-three years old.

While I didn't know it at the time, OSI was a historic company because it had been started by Edwin Link, the inventor of the Link trainer flight simulator. In fact, one of OSI's first employees was Jon Lindbergh, the son of aviator Charles Lindbergh. I later learned from Al McCready of OSI's Santa Barbara office that Jon's famous father used to come out on diving jobs to watch his son work.

I started at OSI in August, building ship fenders in the Chantilly, Virginia, shop at $3.50 per hour. The fenders, which ranged in size from 10 to 16 feet in length, were used to protect large ships from damage as they rubbed up against docks. I persevered with the work, but eventually grew tired of going home covered with particles of polyurethane foam and industrial glue. I wanted to work on diving systems! Every day as I came into the shop, I saw a few diving bells being constructed or repaired. I longed for the day when I could get my hands on such equipment.

After eighteen months of ship fenders, I finally got my chance and started working with 600- and 1,500-foot-capable diving systems.

Commercial deep-diving systems are far more than the hard-hat gear the public sees. They consist of a diving bell (called an SDC, or submersible decompression chamber), an entrance lock (a smaller chamber used on the surface to transfer the divers out of the SDC), a living chamber (where the divers live while under pressure), and a control van (where diving operations are controlled).

Many of the dive system elements are connected so the diver can transfer from the living chamber into the SDC and go underwater and return—all while under pressure at a simulated water depth. The most common diving technique used with these systems is saturation diving, where the diver is kept at a simulated bottom pressure over a series of days (or even weeks). This enables a far more efficient use of time on-site because the diver is not required to decompress back to the surface after each dive. Decompression is a lengthy process whereby the diver is slowly brought back to the surface from the depth at which they have been working. The process can take several hours or days, depending on the depth.

The dive systems I built consisted of mechanical components, stainless steel tubing, electrical wiring, gauges, and lots of valves. The work gave me a good understanding of engineering and a knowledge of general fabrication techniques. It was an experience that would be valuable later as I moved into the underwater vehicle field.

I later relocated to Houston with OSI to continue working with deep-diving systems. But again, it was not what I really wanted to do. I wanted to see such equipment in operation, or maybe become a commercial diver myself. What I really wanted was to become a submersible pilot. If I could get a job driving a mini-submarine, it would be the fulfillment of a lifelong desire.

I left OSI in 1977 and began training as a commercial diver at the Commercial Diving Center (CDC) in Wilmington (Los Angeles), California. The school was interesting enough as I spent over three months in intensive classroom work on physiology, diving equipment, and commercial oil field practices. I also spent a lot of time in the Los Angeles harbor. At times, the water was so murky I was forced to train underwater completely by feel—a common practice in the commercial diving field.

One time we were learning how to use heavy gear in the training tank at CDC. I was tending a Polish diver and didn't lock his Yokohama hat on properly. While he was in the tank doing some valve assembly work, the instructor snorkeled down and cranked up his air supply to full to see how he would react. Because I had not locked down his helmet correctly, it unscrewed from his breastplate and flew off his head. If this had happened in the harbor, he would have drowned for sure. Fortunately, we saw what happened and hauled him out with a canvas dry suit full of water. To drain his suit, we hauled him up by his ankles. I had made a horrible mistake that could have cost him his life, and it never happened again.

I graduated second in my class and started beating the bushes for a job driving submarines. Unfortunately, I didn't have any luck, mostly because the operations companies were in the process of switching over from manned to unmanned submersible vehicles. These remotely operated vehicles (ROVs) were quite different from manned submersibles in that they were much smaller and, above all, unmanned. No one went underwater. That didn't seem like much fun. Where was the fun in working underwater if you didn't even get wet?

During my training at CDC I had made it a point to stay in touch with Charlie Hedgepeth, OSI's operations manager. He had recommended the school I attended and was always helpful in my pursuits. Shortly after graduation from CDC, I received a call from Dick Frisbie, an engineer I had worked with at OSI, who said to me, "We're in the process of buying an unmanned submersible—the SCORPIO—and we need some people to work with the vehicle offshore. Are you interested?"

I wasn't pleased at the prospect of working with a remote vehicle. *What did it look like? How did it work?* I wondered. I asked Dick some questions about the system. Most of his replies were over my head, such as, "The SCORPIO uses a servo-controlled hydraulic propulsion system controlled from the surface via digital multiplexing and high-voltage AC." *What was digital multiplexing?*

"If you want to come on board," Dick continued, "we'll get you down to San Diego for some training. But in July, we will send you to Taiwan. We already have a job for the system in the Formosa Straits harvesting pink coral." I further wondered: *Why would anyone want to harvest coral, of all things? What would Taiwan be like?*

I met Dick at the airport and got a little more information on the job. Given that I'd had no luck finding employment piloting manned submersibles, I figured working with the SCORPIO would be a start, so I took the job.

I headed down to El Cajon in my uncle's 1952 Chevrolet Bel Air and spent the next three months learning all about remote vehicles.

In 1977, the SCORPIO ROV was considered state of the art, in that it represented the latest in remote underwater technology. The vehicle I was going to be working with was the first of a class—a prototype.

A remote vehicle is a platform used to do remote work and inspection on the ocean floor. There are many ways to accomplish these objectives, but typical remote vehicle designs feature a framework (normally aluminum) onto which is mounted the various electrical, electronic, and hydraulic subsystems needed to view the ocean floor and do any work desired by the vehicle operator.

The vehicle, which can range in weight from 100 pounds to several tons, is maneuvered on the ocean floor by a series of propulsion units, commonly termed "thrusters." These thrusters are really propellers specially designed for underwater use and can be electrically or hydraulically powered.

Underwater viewing and navigation are done using video cameras and obstacle avoidance sonars mounted on the frame of the vehicle. The cameras, also specially designed for underwater use, can be fixed in place or mounted on pan-and-tilt units, which allow them to be tilted up or down and rotated to the left or right. This increases their effective field of view. Given the darkness of the ocean below a few hundred feet of depth, the vehicles are also fitted with a series of underwater lights. The vehicle-mounted sonar is used to find objects on the ocean floor and does this by sending out pulses of sound, which are received by the vehicle's onboard receiver.

Work is done underwater using hydraulic manipulators designed to roughly mimic the action of a human arm. These manipulators can be very powerful. Some even have a sense of feel, called "force-reflection," where the vehicle's operator can feel the forces felt by the arm as it works underwater.

The vehicle is outfitted with a submerged electronics bottle containing all the electronics needed to process the commands sent by the surface operator. These commands, controlled by a microprocessor, are sent from the surface to and from the vehicle via an umbilical, a long electrical cable connected to the surface control consoles. Contemporary umbilicals are made up of fiber optics bundles (for transmitting and receiving video and telemetry), copper wiring (for electrical power), and a strength member normally made of Kevlar (an artificial fiber manufactured by DuPont) or steel wire.

The umbilical is linked to the surface control consoles through a junction box. The control console, or operator's workstation, is where the vehicle is operated from. Typically, the console consists of several video monitors that allow the operator (pilot) to view the ocean floor in front of the vehicle. There are also switches for controlling the movement of the video cameras, lights, and other vehicle subsystems. The actual movement of the vehicle is normally done through a hand controller, or joystick, like that used by pilots to fly high-performance aircraft. There are sometimes a series of automatic functions that enable the pilot to make the vehicle hold a desired course or altitude as they drive it over the bottom.

The vehicle's outward appearance can vary from sleek and streamlined to quite awkward—nothing more than a framework holding the electronics bottle, thruster, cameras, and lights. Given that the vehicle's speed over the bottom is usually less than 1 knot and the fact that most of the underwater drag from currents is caused by its long umbilical, streamlining is more for the sake of appearance.

While most remote vehicles today are fairly sophisticated, in the past, many used a considerable number of off-the-shelf components in their design, such as floor polisher motors (for thrusters), drill motors (for driving the hydraulics), home security cameras (for video, mounted in a watertight housing), sailing boat propellers (for propulsion), and scuba tanks (to allow pressure compensation for air-filled components).

While the SCORPIO was new and unproven, there would later be more than sixty such vehicles manufactured. But we had the first one—the prototype—and it gave us plenty of problems. I couldn't believe how much maintenance was required to keep that vehicle running. I figured we were going to end up spending all our time fixing the sub instead of doing any work with it.

It took us three separate attempts at dock and sea trials over a three-month period before we ever got the SCORPIO working in any real sense. On our last trial, we finally managed to get the vehicle down to 2,000 feet, considerably shy of its 3,000-foot-maximum rating. The next thing I knew, I was on a plane headed for Taiwan—my first real job offshore.

5

1977: The First Job

I n October of 1977, I was twenty-six years old and standing on the bow of a Chinese fishing boat, swinging a bamboo pole holding a string of exploding fireworks. It seemed to be an especially dark, mist-filled night in Keelung Harbor, a fishing port on the north end of Formosa.

The stench from raw sewage and dead animals in the harbor drifted across my nostrils. Numerous dilapidated fishing boats were strewn near the dock like forgotten or broken toys. The *Deepsea I*, with an all-Chinese crew (most of whom did not speak English), chugged out toward the exit. Earlier in the week, I had watched with amusement as scrawny teenagers wearing nothing but grease-stained underwear did "prop jobs" donning only small goggles to protect their eyes. I had never been out to sea at all, really, in my life. The closest I had come was during a charter fishing cruise with my father and brother, and I remember not liking it much.

I arrived in Taipei on September 8, 1977, and initially ended up at the Grand Hotel. But instead of resting after such a long journey, my fellow travelers and I immediately began to cruise the various hostess bars in the city. No big surprise there. However, it was not long before myself, Russ Austin, Chuck Collins, and Glenn Tillman were dragged up to Keelung, the country's main shipping port, where our support ship, the *Deepsea I*, was docked. So, in short order, we went from happiness in a wonderful Taipei hotel to the Kodak Hotel in Keelung, where I spent most of my time killing the dozens of roaches crawling on the walls. Russ, the laid-back diver from Seattle, had worked in Taiwan before, doing commercial oil field work, but the rest of us were mystified by what smelled like rotting meat in the local markets and the large mirrors placed in every doorway. That was Keelung.

Just getting on board the ship was an adventure in itself: To even reach the *Deepsea I* required us to climb over nine other dilapidated fishing boats. Our remotely operated vehicle (ROV), SCORPIO I, was already on board, as were many wooden boxes of support equipment.

At best, the ship was 75 feet long, with a steel hull, wooden deck, and fiberglass superstructure. The vehicle and cable reel were mounted on the fantail, just behind our combination berthing area and ROV control room, which was 8 by 8 feet. There was a single hatch going from this cabin directly to the deck on the starboard side (the right side of the ship looking forward). Inside this room were six bunks fitted with foam rubber mattresses, a small storage area, and a large shelf onto which we placed the SCORPIO control box, sonar display, and video monitor. There were also two large air conditioners, which I later discovered dripped water onto the bunks, mine in particular. As a result, my rack was always sopping wet. A 2-by-10-inch wooden plank served as our seat during operations, propped up by one of the bunks and the open hatch. The area was so confined with all our gear that no more than one person could stand up at a time. Even so, our mission was to harvest pink coral using an ROV in deep water. Intact branches were quite valuable, according to Telo Ma, our client, who represented the Deepsea Resources Development Corporation. Many promises were made, and there was a profit-sharing plan suggested that could potentially make us all a lot of money.

The *Deepsea I* had no galley, so we had to eat our meals on deck in the weather, usually sitting on a large cargo hatch forward. There was one toilet and a gravity-fed shower, which was cold water only. I think the ship had a radar, something like what you would see on a small pleasure craft. There was no VHF voice radio because the Chinese government allowed communications only by Morse code. It was a single-screw ship with no bow thruster, and guys slept in the engine room on little shelves fitted with mats. According to our client, the cook was a former embassy chef. However, all I remember getting out of him was rice and fish.

If I had known anything at all about the ocean and been paying attention to the local Chinese newspapers, I would have realized that a powerful typhoon had just passed through the area. As we chugged out to the open sea, I stared at the rows of other fishing boats safely tied up at the pier. But we were going out, swinging poles of fireworks to scare off the evil spirits and guarantee a successful trip. It didn't work. *And it's a fine Ocean Systems day!*

There was no safety drill of any kind, and I had no clue as to whether the ship even had life jackets or rafts on board; I had only noticed a few life rings. Not that it would have made any difference, because one decent wave on the beam of this overloaded ship and, rock ballast or not, we'd capsize and sink like a stone. (What was my mother going to think when she learned I had drowned in the Formosa Straits for no good reason?) A few hours before, I had been killing roaches on the walls of the Kodak Hotel; now, here I was. (This was our second trip out on the *Deepsea I* with the SCORPIO. The events of the first attempt to do ROV operations from the ship had been laughable.)

Once the ship was in the general area of a probable coral bed, we would launch the vehicle using the winch associated with the A-frame mounted aft (toward the rear). The SCORPIO was released using a pelican hook, and the umbilical floated with aluminum Grimsby fishing floats. I would typically be in the berthing, communicating with the deck via the Helle divers' radio. Someone else would be manning the Helle pinger locator system, sending up the slant range and relative bearing angles to the bridge, where Glenn would take the numbers, do the calculations, and plot our position on a paper chart. This system did not work. We continually dragged the SCORPIO over the bottom because the ship did not know where it was, and we didn't know where we were. We had no way to navigate the ship because it didn't have LORAN (long-range navigation) and there was no such thing as a GPS satellite in 1977. Consequently, using the pinger locator and crunching numbers with a calculator was all we had. Sometimes we anchored if the water was shallow enough. But it was obvious that we could not operate in this fashion.

Eventually, there were phone calls made back to Dick Frisbie (our boss and program manager) in Houston, and it was not long before he came over to sort things out. I distinctly remember our entire crew meeting him at the airport in Taipei, holding up cardboard signs that said, "Welcome Stupid American." He was not pleased.

Once Frisbie was in the country, there were several meetings with Deepsea officials. Nothing was really sorted out because our gear was on their ship and we did not have an easy way to get it off. I must give Dick credit; he was tough with the Chinese. However, in the end, we elected to give it one last try to see if we could get some coral and make everyone happy.

During the standoff between Ocean Systems Inc. (OSI) and the Chinese, we ended up in Taipei one night to have dinner and drink. During this excursion it began to rain. Not a little, but a lot. Before long, there were 2 feet of water in the streets, and we learned that a landslide had taken out the single highway linking Taipei to Keelung. So, it was the next day before we made it back on the train.

As we steamed out in the early morning, I drifted in and out of consciousness, feeling something splashing on my face with a regular rhythm. In time, I opened my eyes, rolled over, and noticed it was seawater squirting in from the watertight hatch all the way across the cabin. We seemed to be rolling a lot, with assorted bottles and shoes sliding back and forth across the deck of the cabin, which was already wet. But a single can of Planters Peanuts, with the plastic top chewed through by a rat, was still in place.

Eventually, Glenn crawled out of his rack, threw on some shorts, cautiously cracked the hatch, and looked out at the seas, exclaiming, "Holy shit!" It was bad. The seas were topping 35 feet, and all the ship could do was hold her bow into the wind. It was like a scene from *The Perfect Storm*, where the fishing boat rides up one side of a wave and crashes down the other side. Many of us got seasick. I kept everything inside of me, though I was not feeling well at all. This would be the start of a very long and terrifying day. Going on deck was out of the question.

Before long, there was so much seawater in our cabin that I ended up sleeping with a spare Conrac video monitor and Ametek-Straza sonar display, just to keep them dry. I didn't know enough to be scared. But I should have been. In his SCORPIO trip report, Dick Frisbie, a retired Army captain and

Vietnam veteran, later wrote that he had been in hurricanes on a research vessel and not been as terrified as he was on that first night on the *Deepsea I*. My most enjoyable recollection of the period was sharing a black market can of peaches with Frisbie. It was the most solid food I had consumed in days. That afternoon, the lot of us squatted on top of the bridge and complained profusely to Frisbie. Chuck Collins, a former airdale and, like Dick, a Vietnam veteran, announced, "I didn't sign up for this deal to drown in the fucking ocean on this piece of shit."

As I took it all in, I wondered out loud, "Are all these jobs like this?" The response was rolling eyes, cursing, and muttering about what an idiot I was.

By September 25, the seas had abated enough that we could start diving. We had given up on live boating (not anchoring) because it was obvious that we could not station-keep (maintain our position) given the situation. We didn't even have an acoustic navigation system, something that is deemed a requirement these days. So, we anchored the *Deepsea I* over a likely coral bed and sent the SCORPIO down, with me at the controls.

The SCORPIO vehicle we were using was leased from Ametek-Straza by OSI, my employer; we didn't even own it. But the company intended to buy it, if and when it worked reliably. It had taken us over three months to finish building the ROV and get through some resemblance of sea trials. It worked, but spent much time broken down. This vehicle was serial no. 01, the prototype, never a good thing to deal with. For the time, it was very advanced from a design standpoint. But I didn't really know if it was any good or not—in fact none of us did—because we had never worked with an ROV before and there were few in commercial use at the time. The SCORPIO had an air-filled aluminum tube frame that actually floated in water and could dive as deep as 3,000 feet underwater. However, compared to contemporary ROVs, it was very limited in its capabilities. It only had one camera, two lights, four propulsion units, a frequency-modulated sonar (also built by Ametek), and a single Perry hydraulic five-function manipulator. The whole thing was controlled from the surface through a single console, essentially a large metal Halliburton Zero case, crammed with tons of switches, knobs, and dials. The contraption was temperamental, to say the least, and difficult to repair. And yet surprisingly for a vehicle built in 1977, several subsystems, such as the sonar and thrusters, are still in use today by the US Navy.

From the beginning in San Diego, I was good at piloting the SCORPIO vehicle; I just had a knack for it. I don't know whether this was because of my racing background or simple hand–eye coordination. In other words, I developed an excellent feel for what control inputs using the small joystick resulted in the desired visual movement on the video monitor. I was just good in general at operating machinery and equipment, whether it was the razor-sharp cabbage slicer at my uncles' produce shop, my Zink Formula Vee, or a belt sander to smooth out the edges of the ship fenders I built at OSI in Chantilly, Virginia. But what I didn't have was much experience with the underwater environment. Sure, I had been to diving school, but that was just stomping around on the bottom of Los Angeles harbor. Anybody could do things when the conditions were good, such as on a nice clear sunny day in a small airplane. But it takes real skill and experience to land a Cessna 150 with a forward slip during a stiff crosswind.

My personal records show that the water depth was about 840 feet and that we slowly rolled back and forth 50 miles offshore of the main island. I made it to the rugged bottom without incident among scattered boulders with reasonable visibility. One thing I didn't see was any coral. Russ and I had earlier rigged a steel basket on the bottom of the vehicle to scoop up intact coral branches. But all it did was make the SCORPIO almost impossible to drive.

Suddenly, after about forty-five minutes of flying around, what appeared to be a rope or wire flashed past the front of the vehicle, shining in the beam of the underwater lights. It took me so much by surprise that I instinctively ducked to keep the object from hitting me. We soon realized that the steel anchor wire of the *Deepsea I* had ripped free of the ship's anchor and wound itself around the SCORPIO's electrical umbilical like a snake. It didn't take long for things to go even more wrong.

In less than a minute, I watched helplessly as the vehicle was dragged backward and up over an undersea cliff. I yelled a warning over the Helle communications box as the video screen first began to flicker, then flash on and off as the steel wire sawed slowly into the umbilical. Bits of rock and sediment rained down around the ROV as it was hauled over the outcropping like a piece of expensive junk. As Russ poked his head into the compartment, we both saw the red lights flashing on the control console when the wire rope cut into the 3,000-volt wiring in the umbilical and tripped all the breakers, effectively killing the sub. SCORPIO was finished and there was nothing for me to do but go out on deck and help haul in the umbilical; we had no idea whether or not there was still an underwater vehicle on the end.

Eventually, the anchor wire and SCORPIO freed themselves and we managed to make a reasonably normal dead vehicle recovery. I could see a few scrapes on the side of the sub from the boulders, but otherwise, everything looked reasonably intact. I went back into the berthing compartment and turned on the high voltage, only to hear a bloodcurdling scream from on deck. Dick Frisbie had been casually leaning on the vehicle's frame when I turned it on. I had nearly electrocuted my boss.

After this incident, Dick declared that Deepsea Resources was in violation of their contract because the company had not supplied a suitable support vessel. The job was over. As we headed back to Keelung, we contemplated whether to re-terminate (cut off and rewire) the umbilical because it had been cut though about halfway by the anchor wire.

Once we reached Keelung, our problem was that there was no way for us to offload our equipment because the ship was moored on the outside of a half-dozen fishing boats. Deepsea Resources refused to move the vessel. What followed was a week of uncertainty, with all of us simply hanging around in Taipei; we could not even leave the country because Deepsea refused to complete the necessary immigration paperwork. But every day, Russ and his Chinese girlfriend made the trip to Keelung to check on our equipment. One day, though, the ship was *gone*, along with the SCORPIO system. We left. From what I gather, Telo Ma had somehow figured out how to do a re-termination from watching the procedure at Ametek. According to Deepsea, they then operated a vehicle they did not even own and lost it at sea. The truth remains a mystery and the State Department was subsequently notified that advanced underwater technology had been stolen by the Chinese. I think it was officially labeled an "act of piracy." Many months later, the SCORPIO's topside gear, minus the ROV, was returned to Long Beach, California. Thus ended the saga of SCORPIO I.

By the end of the Taiwan job, I had accumulated twenty hours of piloting time over seventeen dives, with my deepest dive to 2,039 feet of seawater. I thought I was an expert. I could not have been more wrong.

6

1978: The Oil Patch

After the debacle in Taiwan, I worried that I might not even have a job anymore at Ocean Systems. But the company soon put me back to work in the shop outfitting an underwater pipeline alignment and repair system. This was another area in which OSI was a true pioneer. The company designed and fabricated a pipeline alignment vehicle that could be positioned far underwater over a damaged pipeline. The system even had a docking ring for one of the ADS-4 (advanced diving system) diving bells. Once the bell was attached to the rig, the divers exited the bell and blew the interior dry with either air or mixed gas, depending on the depth. The alignment frame and associated dry chamber was huge, easily over 40 feet long, and weighed many tons.

Once the interior was pressurized, the divers could cut out the damaged section of pipe, insert a new piece, and weld it into place. The massive hydraulic rams on either side of the work chamber were used to align both ends of the pipeline so the divers could replace the damaged section. Once they were done, they even had the ability to x-ray the welds to make sure all was done to offshore standards. Then they would pack the joints with tar and cover the connections with sheet metal, flood the chamber, ingress into the diving bell, and be on their way back to the surface. It was a marvel of engineering and operations. After working on the system for several months, OSI started thinking about buying some new remotely operated vehicles (ROVs) from a little-known Canadian company, International Submarine Engineering (ISE), formed by James McFarlane, a former Canadian Navy submarine commander.

In 1978, the business of remote vehicles was still very much in its infancy. At the time, there were probably less than fifty commercial remote controlled underwater vehicles in existence, the most popular a model called the RCV-225 (RCV stands for remotely controlled vehicle). The RCV-225 was a flying eyeball of sorts, not unlike the Jason Junior vehicle built by the Woods Hole Oceanographic Institution. While it could inspect objects underwater, it was unable to do any real work because it had neither the power nor the manipulators for such activity. The RCV-225, which was about the size of a basketball and weighed about 150 pounds, even used Thimble Drome model airplane propellers to thrust itself through the water. It was very good for simple inspection tasks, but useless for anything else because it simply did not have enough power.

At about this time, ISE was starting to build some larger remote vehicles capable of heavy work underwater. I didn't know much about ISE except that the company was made up of a bunch of engineers with considerable experience in manned submersibles of the Pisces class. The Pisces manned submersibles were depth rated for anywhere from 3,000 to 6,000 feet, depending on the model. They were built in Canada and had amassed an excellent reputation for use in oil field, scientific, and cable burial applications. But now many of the engineers who had built these submersibles and some of the pilots who had driven them were building remote vehicles in Vancouver, British Columbia.

They built a new remote vehicle that OSI was considering buying called TROV (tethered remotely operated vehicle) S-4 (for the fourth submersible built). It was large, powerful, and featured two manipulators. OSI bought not only that submersible but four more as well to corner the market on unmanned work vehicles. It was a visionary move because the company saw the future of remote vehicles with respect to their evolution. The problem was that such equipment was still just being accepted as a basic inspection tool in the oil fields.

OSI forged ahead anyway by purchasing three TROV heavy work vehicles and two lighter TRECs (tethered remotely operated cameras) for basic inspection and light manipulative work. The TRECs were considerably different from the TROVs in that they weighed only about 500 pounds and had one small arm. The propulsion system used by the TREC was based on a series of floor polisher motors (yes, real floor polisher motors) mounted in aluminum housings pressurized with air from a standard scuba bottle. In fact, the vehicle didn't even have much of a telemetry control system at all. The compass display for the TREC consisted of a Panasonic security camera pointed at a diver's compass.

In 1978, OSI began to use TROV S-4 for oil field work in the Gulf of Mexico, the North Sea, and the Mediterranean. Our first Gulf job was done from an all-aluminum work boat called the MV *Sea Rambler*. Why OSI thought it was smart to put us on an aluminum boat was beyond all of us because it greatly complicated mobilization. Maybe they got a good day rate. Consequently, all our gear had to be chained down using pad eyes welded to the deck.

By today's standards, our launch and recovery system (LARS) was laughable. To launch the TROV, we were using a surplus open bottom bell davit that weighed considerably less than our ROV. The only way we could rotate the vehicle over the side was by using a couple of air tuggers (air-powered winches). We lifted the vehicle the same way. Every time we picked up the vehicle, the davits boom flexed like a leaf spring, and it required constant repairs. In addition, we had a Bolstad diving compressor as a counterweight to power our multiple air tuggers. While hauling the heavy TROV into the air, the skids of the LARS slapped on and off the deck, making a sound like a heavy machine gun.

As usual, there was no place for us to sleep, and we crashed on mattresses strewn about the deck of the galley. Our captain was Cajun and spoke with such a heavy accent that we literally could not understand what he was saying. While he knew how to handle his boat, he was totally ignorant about how to station-keep using an acoustic navigation system, a new addition to our efforts. The man was a very scrawny, malnourished individual who liked to sit back on his chair on the bridge, chain-smoke Picayune cigarettes, and steer the boat with his bare feet. But he did the job. While I think LORAN (long-range navigation) was around, this was long before the advent of satellite navigation, and most boat captains navigated by dead reckoning and reading the oil rig numbers as they cruised the Gulf.

This was our first time using an underwater tracking system, and we were outfitted with a Honeywell system called the RS-7. All it did was give us range and bearing on a circular cathode ray screen, which was fine with us because that was all we needed. But in the late 1970s, many mud boat captains could not understand how to use this system. They simply could not visualize how it represented the vehicle's relative position. Eventually, we relented and used the old Canadian trick of floating the vehicle's umbilical on the surface so the captain could see it. This he could manage.

Our job was to inspect a pipeline with TROV after a recent installation by a lay barge. As with the incident with Nekton Alpha described in the first chapter, the surveyor's buoy lines were a familiar hazard to any ROV. All went well for a while, until we inevitably snagged a surveyor's ¼-inch poly buoy line around TROV's umbilical. That tiny line sawed halfway through the umbilical and we had to do a dead vehicle recovery, which means we had no control of the thrusters or any way to maneuver TROV and had to recover it by its now damaged cable.

Our next problem was that we were out of the epoxy resin needed to repair the cable. So, we improvised and did an in-line splice of the cable using copious amounts of solder and electrical tape. To keep it from flexing underwater, we tied a section of lumber to the side of the splice and completed the rest of the job in this manner. At one point, after we damaged the launching davit (I had forgotten to remove a safety pin) and had to berth the *Sea Rambler* way up in the bayou in what appeared to be little more than a creek (there was no dock), we called up a local welder and he parked his truck on a nearby road, ran his welding cables down to the deck, and repaired the davit. The whole time it was miserably hot and humid, and we were all eaten alive by the massive Louisiana mosquitoes. But we got the job done.

After initial operations with TROV, the next thing I knew I was on a Norwegian supply boat headed for the Ekofisk oil field in the North Sea. Our job was to inspect the interior and exterior of a series of Phillips Petroleum platforms using a TREC vehicle and take cathodic protection readings along the way. The cathodic protection readings would give the engineers an idea as to the level of corrosion on the oil rigs.

What it's like to work in the North Sea depends largely on the nationality of your support vessel and your method of transport to and from the oil fields.

Our vessel, the MV *Ibis 7*, was a Norwegian supply ship 175 feet in length. The fact that it was Norwegian meant two things: Norwegian food (cold cuts for dinner and lots of fish) and, most importantly, no beer or liquor. I lived on cheese and crackers for weeks.

In the case of the *Ibis 7*, there were insufficient accommodations for our eight-man ROV crew (four men to each twelve-hour shift—midnight to noon and noon to midnight). As a result, we had to cram our team (made up of four homesick Americans, three Brits, and one Norwegian) into a large community cabin stuffed deep inside the ship. The fact that the room was designed solely to house oil field workers for twelve hours (as they were transported to and from the oil field) didn't bother our employer. Of course, they weren't on the ship. The problem was that the room had no air circulation and was located directly behind the ship's sewage tanks. In addition, because the cabin was designed for temporary accommodation, there were no lockers to store our clothing. Consequently, the interior of the cabin ended up being one large closet of sorts, with the clothing and personal effects of one person easily mixed with those of another (mostly on the floor). To make matters worse, the two electrical generators used to power our TREC remote vehicle were welded directly to the deck right above our heads. No one had considered the noise they would create as they droned twenty-four hours a day.

The TREC-5 console used to operate the vehicle was simple in comparison to the ones I would later operate: just two small black control sticks, some switches for lights, assorted electrical controls, and two 9-inch video monitors. Only one monitor was used to observe the bottom; the other showed an image of a diver's compass and an LED depth readout.

The oil platform we were tied up to was a twelve-legged steel structure called Bravo platform sitting in 230 feet of water. Despite the shallow water depth, the overall height of Bravo was probably equal to that of the Washington Monument, and there were some formidable operational challenges related to the job.

To start with, we could inspect certain areas of the platform only at specified times of the day due to the tidal changes in the currents. Invariably, I had to make sure we were working into the current so the TREC's small ½-inch umbilical would be blown *out of* and not *into* the rig. Second, the rig and the entire surrounding area was a veritable junkyard, with piles of broken machinery, steel structures, and pipes littering the bottom around the rig. There were sections of steel structure hanging off parts of the rig—all a result of people dumping their trash over the side of the platform. In addition, given the boredom associated with life on Bravo platform, there were bound to be scores of monofilament fishing lines, lead weights, and hooks all over the rig's structures—just waiting to entangle the TREC's electrical umbilical. I was required to inspect both the inside and outside of the rig, even though we really didn't know the design of the structure under the waterline. Many times, we discovered the engineering drawings to be inconsistent with actual structures. I didn't even have a navigation system to guide me through the maze of horizontal, vertical, and diagonal support members—it had to all be done by sight, and I'd better not get lost because if I took one wrong turn, the vehicle would get stuck on the rig for sure.

To set up the *Ibis 7* for ROV inspections, a bow anchor was put out and the stern tied off to a corner of the rig. This was normally done by the deckhands, but for some reason, as the supervisor, the responsibility fell to me as I soundly slept in our berthing.

I felt a tug on my arm as a voice in the darkness said, "Can you help moor the ship?"

"Can't the ship's crew do it?"

The voice in the darkness replied, "No, they're all off shift and sleeping and we need someone to secure the stern line to the flare stack."

Cursing and moaning, I rolled out of my rack and pulled on my insulated coveralls and Red Wing boots. The next thing I knew, I was in a small Avon rigid-hulled inflatable boat holding onto the end of the mooring lines as the *Ibis 7* held station in the darkness.

The ship's hands managed to nudge and hold the bow of the boat against one barnacle-encrusted leg of the structure as I climbed up, dragging the spliced end of the mooring line with me onto a very narrow set of ladder rungs. I was now hanging onto the side of the platform leg, without a life jacket, hard hat, or gloves, with my left hand on the rung and my right hand holding the mooring line.

Above me the flame from the flare stack hissed loudly in the early-morning hours. It was freezing cold, except for the radiant heat blasting from the 30-foot flame of gas spewing from the stack. While we had flashlights, they really weren't needed because the entire area was lit up with red and yellow light from the burning gas.

"I need more slack!" I screamed as the ship's hands tried to feed me more of the hawser.

It took every bit of my strength to wrap the line around the strut, and they managed to grasp the end and secure it with a rusty shackle. It was snowing more heavily now and the time was just approaching 3:00 a.m. *What a life*, I thought as I carefully managed to step off the rungs and collapsed into the boat. Even as a young man, I was physically exhausted. But we still had a dive to do so I couldn't go back to bed.

It was 4:00 a.m. as we readied the small TREC vehicle for launch into the frigid waters of the North Sea. Surprisingly, it was already daylight and snowing lightly as the *Ibis 7* tugged against the single stern line attached to a corner of the rig that I had just installed.

Our procedure for getting the TREC onto the rig was simple. We would first launch the vehicle using a small crane welded to the starboard side of the ship. The launching system was nothing fancy, just an old lifeboat davit we found in a local Norwegian marine junkyard. Government officials, who had earlier inspected the crane, eventually told us it was unsafe, and we were prohibited from using it. But by then, fortunately, the job was over.

Once the vehicle was in the water, the deck crew would then guide me toward a corner of the rig until I had made visual contact with some part of the platform. From then on it was just a case of keeping sight of the structure underwater and making sure the TREC stayed clear of any debris.

Squinting at the TREC's single video monitor, I could just make out one corner of a large riser (a vertically running section of pipe) jutting out of the platform as the TREC bounced up and down in the swells. Although it was the TREC and not me that was underwater, I felt as though it *was* me and not an inanimate piece of equipment getting bashed around in the seas. The TREC's single camera was black-and-white, so there was no sensation of color as I studied the image of a heavily encrusted section of steel platform. In fact, the marine growth on the rigs almost appeared to glow as it reflected the light from the TREC's large spotlight.

I was situated at the TREC console wearing blue insulated coveralls and Red Wing boots—they're laceless, so if you end up in the water you can kick them off your feet. (I always wondered about that, though. What's the point? The water is so cold you would only last about thirty minutes anyway, so why worry about swimming?) Outside, I heard one of our generators roaring on deck as interference from the TREC's propulsion units made whirring sounds in my headset against a background of static.

"OK, you look like you're in pretty good shape—go ahead and take it down," instructed a British crew member.

I took the TREC's left control stick and tilted it slightly to one side. The whirring sound picked up in intensity—we were going down. On the video monitor, I watched my progress as the vertical pipe traveled upward across the monitor. Every so often I saw large clamps held on with rusted bolts jutting out from the corner of the rig. *Better stay clear of those*, I thought. The tidal conditions were pretty good, because all I had to do was keep a little forward thrust on the vehicle and stay a consistent distance from the side of the rig.

To drive a remote vehicle, you've got to mentally transport yourself into the environment. I wasn't on the deck of the *Ibis 7* driving the vehicle. My thought processes, everything that made up my mind, were underwater with the vehicle. In fact, once you look around underwater, you are sometimes able to mentally construct the environment around the vehicle. It's sort of like creating a mental virtual reality. While all you see is displayed on the video monitor, you know the relative positioning of objects around the vehicle, because you've seen them and have mentally kept track of their location. You can almost hear a *thunk* as the TREC bumps up against the side of the rig, and the whirring headphone interference from the vehicle's floor polisher thrusters helps you to gauge the amount of thrust you're using and in what direction it is going. But it's something you can only do in a visually structured environment. On a featureless bottom, you are lost in a darkened void.

I was finally at a point where I needed to do some work—in front of a heavily corroded anode, a sacrificial chunk of metal designed to lessen the corrosion on the steel platform. Gently, I tried to nudge the vehicle's sharp metal probe against the anode. To take the cathodic protection reading, I had to touch a sharpened metal probe mounted on the front of the vehicle against the anode. But I couldn't seem to get a proper reading with all of the marine growth on the anode. Backing the TREC up about 10 feet or so, I took a running start at driving the probe through the anode's heavy coating of growth. The TREC banged into the probe as I used the vehicle's thrusters to maneuver the probe around a bit into the abode. Finally, the instrument's LED readout counted and I got a decent number. On to the next one . . .

The scene underwater was unique. Two hundred feet below the surface, there was still some ambient light, but not much because the shadow of the towering Bravo platform darkened the ocean floor below the rig. The water was frigid and particles of debris swirled around the TREC as its small thrusters strained to guide the vehicle around the rig's exterior. The actual structure of the platform was composed of a vast network of steel tubes, each one measuring approximately 2 feet in diameter. The symmetry of the interior of the platform was interrupted by a series of conductors, vertically oriented steel tubes running from the surface to the bottom through the center of the rig. Because the size of the rig increased with the water depth, pieces of rusted pipe and reams of fishing line decorated the submerged steel framework—hanging like discarded Christmas ornaments on a barnacle-encrusted tree. North Sea fish also thrived throughout the environment, following the lights of the TREC as it continued its inspection tasks. Near the surface, the marine growth was very heavy. But farther underwater, the thickness of the barnacles lessened, giving way to a rust-speckled coating of orange primer.

The TREC was an ungainly looking craft, made up of a fragile aluminum framework topped by a yellow flotation package. On the front of the package were the letters "REC"—the "T" had long been ripped off the vehicle's front following several altercations with Norwegian steel. Numerous gashes were apparent on the vehicle's top. On the front was a single black-and-white video camera, one modified diver's spotlight, and a single clumsy manipulator. Evinrude outboard motor propellers rotated inside of the TREC's thruster nozzles, completing the picture of an underwater vehicle thrown together from off-the-shelf bits and pieces. Trailing behind the TREC was an orange electrical umbilical, marked every 10 feet or so with black electrical tape, placed there for the specific purpose of informing the TREC's pilot of the amount of umbilical remaining behind the vehicle as he explored the interior of Bravo platform. In between the black tape markings were numerous gashes in the cable's outer jacket, indications of the abuse the umbilical had seen against other platforms—tug-of-wars with a rig not always won.

Now I was inside Bravo, trying to examine some of the diagonal brac-
ing around the conductors. The TREC was probably about halfway inside
of the rig's structure—about as far as I could get from this side. But the
TREC was now handling strangely. For some reason the vehicle kept sinking. It was trimmed out before we ever put it in the water, so why should it
be heavy now?

Suddenly, a billowing dark cloud of sediment began to fill the screen. It almost looked as though it was being pumped in from somewhere.

"I've got big problems . . . There's something being pumped into the water . . . Better get out of here!" I yelled.

"What do you mean? It looks OK out here," replied the deck crew.

"I'm telling you, they're putting something into the water. I've got to get
out now."

Quickly, I rotated the TREC around, sighted my umbilical, and followed
it out of the rig. The current was rapidly moving the dense cloud of sediment toward the vehicle. If the TREC got caught up inside the cloud, I
would lose visual contact with the rig. I wouldn't know which way to turn or
whether I should go up or down to stay clear of the rig. The TREC would get
stuck for sure.

"Start pulling in cable," I told the deck.

I needed to keep the loop of cable behind the vehicle as small as possible so
it wouldn't snag on something on my way out of Bravo. Finally, the TREC was
clear and the vehicle shot through a wide opening between a diagonal brace
and a vertical strut. After driving straight out of the rig for a second, I rapidly
swung the TREC around in a 180-degree turn, then looked back.

As the deck crew laboriously hauled in cable on deck, I observed the last of
the TREC's orange umbilical as it finally fell clear of the platform. A few seconds later the TREC was completely blinded by the dark cloud of sediment
emanating from the center of Bravo platform. *That was close.* (I found out
later that the rig had started pumping drilling mud into the water. The platform crew hadn't known we were operating under the rig and the extent that
the mud would interfere with our inspection. I'm not sure they would have
cared all that much anyway. What's a $150,000 remote vehicle in comparison
with a multimillion-dollar offshore oil platform?)

After the TREC was recovered, I looked inside its vertical thruster. I thought
something might be wrong with it because of the vehicle's peculiar behavior

underwater. As it turned out, there was nothing so strange about it after I extracted a large steel shackle weighing approximately 5 pounds from the TREC's vertical thruster, along with about 30 feet of fishing line. Just one oil field worker's idea of a fishing weight.

We spent around six months in the Ekofisk field, inspecting platforms ranging in size from a simple eight-legged structure to the main Ekofisk complex, itself over a mile in length. The center of the main complex consisted of a multimillion-gallon concrete oil tank, which we spent several weeks inspecting. At nighttime, the Ekofisk field developed an almost medieval character as the horizon was illuminated by the massive natural gas flames swirling around the Norwegian production platforms. At times, while working under the flare stacks, the length of the natural gas flame would suddenly multiply in intensity, making the deck of the *Ibis 7* as sweltering as the hottest Texas summer day. It seemed hard to believe that such places even existed, that such activity was going on in such an inhospitable place, both on the surface and underwater.

Because I was the supervisor, I was required to send in daily reports to our sponsor company in Stavanger, Norway, TERO A/S, an offshore maintenance firm based at the airport. This was in the days before email, satellite phones, and the internet, so it was a complicated task, especially from a ship at sea. The first thing I had to do was take our handwritten dive log and transpose it into text using a telex machine on the bridge. This was hard for me because at the time I didn't type very well and had to make numerous corrections. Once the report was completed, it would be converted into a long paper tape punctured by hundreds of small holes, which coded the report. This tape was like the ticker tape used to transmit stock price information. Once I had the report on tape, I had to use the VHF radio to contact Rogaland Radio on the beach to get a line on which to transmit the report.

When I thought they had assigned me a clear line, I would feed the paper tape back into the telex machine and send it off. This usually worked without any problems unless you were trying to use a line already taken by another company. This happened to me once as I was sending my daily report. Suddenly it stopped, and I watched with some amusement as someone typed back to me, "FUCK off my line!!! INTERSUB." That was the state of technology at the time. To call home, I had to follow a similar procedure using ship-to-shore radio, which Rogaland Radio would plug into the regular international phone system. The quality of these calls was horrible, and my mother sounded as though she was speaking through a long tunnel.

It was finally my turn to get off the *Ibis 7* and head back to Stavanger for a one-week stint on the beach—working in the office. Even that simple act required a lot of logistics. First, I had to pack up all my belongings and jump from the *Ibis 7* to another similar workboat (this one was going to a nearby oil platform). Unfortunately, by the time we arrived at the rig, the weather had deteriorated to the point where it was too foggy for the helicopters to find the platform. But not to worry, because one of the oil rigs had a helicopter already on it, waiting to leave. With the helmsman holding the supply boat just next to the platform (Alpha platform), their crane operator lowered a personnel basket to the deck of the ship, into which I tossed my belongings and onto which I hung like a marked man. After being lowered onto the rig, I dragged my luggage to the uppermost part of the platform (up several flights of stairs), where I picked up a survival suit and joined the throng of waiting oil field workers. Several hours later, I was finally stuffed inside a shaking Sikorsky S-61 helicopter on my way to Stavanger. It was so foggy that I couldn't see the boat that had just dropped me off. After carefully checking the fuel for contamination, the pilot took off straight up, narrowly missing a boom crane, climbed above the weather, and headed straight back to Stavanger. Later, I learned that they typically lost one of these helicopters per year, many times with no survivors. On the back of the seat in front of me was a small placard informing me that the helicopter I was riding in was "the safest mode of air transportation in the world." *If that was the case*, I wondered, *then why was I wearing a survival suit designed to help keep me alive in the event of a crash?* I glanced out of the chopper's window at the white-speckled icy North Sea. Nope—I wouldn't last long down there—even with a survival suit. Of course, once the helicopter crashed and rolled over, I wouldn't need a survival suit anyway.

Once back in the States, I started freelancing on various ROV operations, in part thanks to ISE, who was always happy to refer me to their customers as an ROV pilot experienced with both the TREC and TROV systems. My first job as an independent contractor (a status that was, and is still, common) was to support Subsea International, who had two TREC vehicles installed on a diving support vessel (DSV) called the *Arctic Seal*. They were stationed off the coast of Campeche, Mexico, assisting in the closure of a massive oil well blowout, Ixtoc I.

Involved in this effort were an assortment of companies and nationalities, such as diving contractor Diavaz and Pemex, the Mexican state-owned petroleum company. At the time, Ixtoc I was the largest oil spill in history, pumping out over 30,000 barrels of crude per day into the Gulf of Mexico. The semi-submersible rig was forcibly pulled off the drilling site following the blowout, resulting in massive damage to the well and lots of fire. The well was leaking not only crude oil but natural gas as well, which the ship's crew kept burning by firing parachute flares into the leak. As a result, what we were faced with was a layer of crude oil on the surface, easily over a foot thick, topped by a plume of gas burning in a torrent in the center of the leak.

I flew down to Campeche via Mexico City and for some reason was put in charge of ROV operations. I also coordinated our work with famed oil well firefighter Paul Neal "Red" Adair. The big problem we had was that we simply could not launch either one of the ROVs and make it down to the blowout preventer (BOP) to monitor the leak, which was our job. There was way too much wreckage to navigate around, so we would dive one of the TRECs and have it guided to the wellhead by a diver. Once there, he would tie the vehicle to the BOP stack by a 20-foot piece of line so we would not get stuck. But that was only part of the problem because even if we got the ROV to the wellhead, the crude oil floating on the surface coated the lens of the vehicle's camera, making it impossible to see anything underwater.

This is where our crew developed an ingenious solution before I even arrived on-site. What they did was take a standard round paintbrush, drill a small hole in the center, and insert a small piece of plastic tubing. The other end of the tube was connected to a plastic bottle of dishwashing soap, which they stuck into the gripper of the TREC's one manipulator. The brush was mounted so that we could pan and tilt the camera and clean it off with the brush. Once the vehicle was tied off to the stack, we would rotate the camera around and squeeze the bottle with the gripper, which would squirt out enough soap to clean off the camera lens. It was crude, but effective.

In total, I spent about a month on the *Arctic Seal* using the two TREC ROVs to inspect the leaking BOP when needed. What this meant was that for the most part, the vehicle was not even turned on as it drifted around the stack, confined by the lanyard. The rest of the time, our crew watched movies, laid in the sun on the helicopter deck (giving us excellent suntans), and exercised on the bicycle on the bridge until the first mate kicked us out. I think we only actually worked four days the whole time. We also kept a close watch on all our gear because the Diavaz divers were stealing everything not bolted down. Ultimately, with Red Adair's guidance, Pemex managed to shut down the well, haul up the battered BOP stack, and save the world. After returning home to Oakland, California, with my sun-burnished skin, I was invariably asked, "Have you been on vacation?" Nope, I've been working.

After Ixtoc I continued to freelance, using my grandmother's house as a base of operations and occasionally working at the Arakelian Brothers produce shop on Fourth Street in Oakland to keep some money coming in. Using all of the produce-processing contraptions made by my Uncle Charlie felt about as dangerous as working offshore because his carrot slicer, as well as the cabbage slicers, could easily take off several fingers. I also helped the brothers keep their assembly line going, and they were impressed by my mechanical knowledge.

In 1981 I was offered a full-time job by Ocean Search, formerly Alcoa Marine, to help with the SCARAB II cable burial and maintenance vehicle they operated for AT&T and Transoceanic Cable Ships, based in Dundalk, Maryland. I relocated from California to Maryland, shipping all my stuff along the way, and rented a small one-bedroom condo just across the bridge from the city dock.

There were two SCARAB vehicles, one maintained and operated by AT&T and the other operated by Cable & Wireless (C&W) of England. While the two vehicles were nearly identical, they were not exactly alike, and fixes for nagging problems were traded between us and the Brits.

The two SCARAB ROVs, I and II, were built by Ametek-Straza, the same company that built the SCORPIO vehicle stolen and then lost in Taiwan. However, they were purpose-built for cable repair work, were much heavier than the SCORPIO, and could dive to 6,000 feet underwater. These were the first two underwater vehicles specifically designed to maintain the existing network of submarine telephone cables that linked the continents. Bell Laboratories (then a subsidiary of AT&T) had a significant role in developing the technology that made it possible for the unmanned submersible to locate, track, and repair underwater communications cables.

What most people don't realize is that even today, most international communications, even military, are relayed using submarine telephone cables and not satellites. When I started working in the cable repair business, all these submerged cables were coaxial (not optical fiber), and they were of varying classes, depending on the application, ranging from double-armored to single-armored to lightweight cables, which had no armor at all. The armor was made from the plow steel we still use on sonar tow cables and consisted of layers of steel wires wrapped around the cable to protect it on the seafloor. The first transatlantic cable was installed in the 1850s from Valentia Island off the west coast of Ireland to Bay of Bulls, Trinity Bay, Newfoundland. The first communications occurred on August 16, 1858, but the line speed was poor, and efforts to improve it caused the cable to fail after three weeks. These first cables could transmit telegraph signals but not voice signals. In fact, remnants of these cables are still dredged up today during the installation of newer fiber-optic cables. They were nothing but a piece of copper coaxial cable protected by jute and manila rope.

These days, there are thousands of fiber-optic cables around the world, each one with repeaters at specific intervals to boost the signal. Some of the developing nations are even now receiving their first submarine cables to put them in touch with the world via voice, data, and internet.

With the development and fielding of the two SCARAB vehicles, all of that was in the future because it was our job to help maintain the existing coaxial cables using the ROVs from a variety of platforms, such as the CS *Long Lines* (AT&T), CCGS *John Cabot* (Canadian Coast Guard and Teleglobe), the CS *Mercury* (C&W), and the CS *Leon Thevenin* (France Telecomm). Ocean Search of Lanham, Maryland, was contracted to AT&T to operate and maintain the SCARAB II vehicle on their behalf in 1981. This was my new employer.

I started off with Ocean Search doing cable work from the CS *Long Lines* in the North Atlantic. These were sea trials to check out many new upgrades integrated by both Ametek-Straza and Bell Laboratories. For that reason, we had both SCARAB vehicles on board, the AT&T and C&W systems. Unfortunately for us, they started off with the British vehicle (SCARAB I). We had so many problems that once a fix was discovered on SCARAB I, the change was replicated on our vehicle, SCARAB II. While in theory this may have seemed like a logical procedure, in reality the AT&T vehicle was not given nearly as good a sea trial as her British counterpart. Consequently, even though our vehicle had been deep, we didn't have nearly as much confidence in it as we had in the C&W system.

After the sea trials, we were sent to Japan to do more cable work, using the KDD *Maru* as a platform. Unfortunately, we snagged a fishing net in our four axial thrusters and they burned up. Why the electrical circuit breakers didn't trip was a mystery we never solved. The four high-voltage propulsion units were oil-filled and ran on about 3,000 volts AC. They were custom made for SCARAB by Franklin, and as a result, the job was put on hold until we could get the motors repaired by the manufacturer.

By May of 1985, we had completed the KDD job and were back at work in the North Atlantic, doing a submarine cable inspection and repair off Providence, Rhode Island, from the CCGS *John Cabot*, a combination ice breaker and cable repair ship operated by the Canadian Coast Guard out of St. John's, Newfoundland. Little did we know that efforts by Sikh terrorists would modify our schedule through their efforts to bring down two Air India 747 jumbo jets the next month. This job was scheduled to last about ten days. I did not see home again for about six months.

7

1985: Air India Flight 182

I was relaxing in the officers' lounge when I first saw the news reports on the Air India Flight 182 disaster. I wondered at the time whether we might be called in to assist in the crash investigation because the water depths mentioned at the crash site were within SCARAB's capabilities. But I dismissed that possibility since we were doing cable work and SCARAB wasn't outfitted for such a salvage job. (The fact that 329 mostly Canadian citizens had just been killed in the crash and we were operating from a Canadian government ship completely slipped my mind.)

However, it did make sense for us to be involved. The British team, as capable as we all knew they were, didn't have any of the specialized equipment or experience to conduct a major salvage operation in waters over a mile deep. Granted, they later did a good job of finding the flight data and voice cockpit recorders, but there is a big difference between simply carrying an object from the ocean floor to the surface and doing a complicated four-point rigging job over several dives and raising wreckage weighing several thousand pounds.

The reality is that the first victims of the Air India disaster were not Indian, or even Canadian: they were Japanese. These were the two baggage handlers who were killed while transferring luggage destined for Air India Flight 301. Their names were Hideo Asano and Hideharu Koda, and the bomb destined for that plane blew up in their faces at Narita Airport. It was a diabolical plan to bring down not one, but two Air India jetliners out of the sky. The only thing that saved the passengers of Air India Flight 301 was that the monsters who conceived their plan didn't know that Japan did not recognize daylight savings time. As a result, the bomb exploded on the tarmac, instead of in the airliner.

The last moments of Air India Flight 182 are not documented anywhere because there were no survivors to recount the events leading up to the disaster. The aircraft's flight data recorder, later recovered by the British SCARAB I submersible, provided precious few clues as to the cause of the crash. Officials studying the contents of the 747's cockpit voice recorder discovered only normal sounds until the last fifteen seconds of the flight. "Then," said R. V. Kunzman, a senior Boeing aircraft engineer, "there was a sudden increase in sounds and the tape abruptly ended." In addition, investigation experts examined at least sixty-four types of information from the jumbo jet's data recorder but got "more questions than answers."

The basic facts surrounding the loss of the aircraft, named *Emperor Kanishka*, were simple. The 747 jetliner was cruising on a flight from Montreal, Canada, to London, England, carrying 329 passengers and crew members made up of mostly Canadian men, women, and children of Indian descent. One hundred ten miles off the coast of Ireland, the Boeing jetliner vanished between radar sweeps at Shannon control, in Ireland, for no apparent reason. A Royal Air Force Nimrod anti-submarine warfare aircraft immediately scrambled from Kinloss, Scotland, out to the last known location of the jumbo jet, sighting drifting wreckage and over 100 floating bodies. The remains of *Kanishka*, now consisting of perhaps thousands of wreckage fragments, rested on the bottom of the Atlantic Ocean in waters over a mile deep.

Had an observer been present during the downing of Air India Flight 182, they would have seen the 747 jumbo jet cruising peacefully at 31,000 feet high above the Atlantic Ocean. There would have been an abrupt flash of light from the airliner's forward luggage compartment, as streams of vapor spewed from the aircraft's now ruptured fuselage. The pilot's attempt to trim the nose of the aircraft down would be to no avail as the aircraft pitched up and stalled. (The jack trim screw was later found in the full nose-down position.) The observer would have then seen the *Kanishka* begin a death spiral toward the hard Atlantic Ocean below.

Because of the potential political nature of the disaster and the suspicion that a Sikh terrorist bomb had been the origin, it was imperative that the cause of the accident be determined. If it was the result of a terrorist act, then those guilty had to be found and punished. If, on the other hand, it was related to a mechanical fault on the airliner, the worldwide fleet of 747 aircraft could potentially be grounded for safety reasons.

The first thing that happened was the US Navy dispatched their PLS-20 towed pinger locator system off the Irish coast to pinpoint the general location of the aircraft's flight data recorder and cockpit voice recorder, which were both fitted with 37.5 kHz acoustic beacons. This they were able to do after searching the general area in the 6,200-foot water depths.

Beyond locating the data recorders, the plan the Canadian government and the US Navy developed was as follows: We had two cable ships working on opposite sides of the Atlantic doing routine underwater cable repair work with SCARAB I and II. The British vehicle (SCARAB I) was installed on board the French vessel CS *Léon Thévenin* and was manned by a team from C&W. SCARAB II, operated by us and installed on board the CCGS *John Cabot*, was concluding repair work on TAT-4 (transatlantic telephone cable no. 4) off Rhode Island. Both ships were pressed into service to investigate the accident; SCARAB I would locate the flight data and voice cockpit recorders, with SCARAB II mapping the debris field and recovering larger sections of wreckage. The American work was supported in part by the US Navy's Supervisor of Salvage and Diving, which supplied heavy lift winches and personnel to do the actual lifting of debris from the seafloor. This specialized equipment was installed on board the German vessel *Kreuzturm*. The SCARAB II vehicle would attach long cables to the wreckage and lift smaller bits of debris directly to the surface.

The French vessel and British team successfully recovered both data recorders in less than a week as we steamed across the ocean to join the investigation. However, the recorders were found to contain little usable information, and most of the public and media interest in the effort ceased after their recovery.

We arrived in Cork, Ireland, on July 9 following a stopover in St. John's, Newfoundland, to pick up some special recovery tools, a 10,000-foot-long Kevlar recovery line, and spare parts for the SCARAB. Our weeklong journey had been full of anticipation at the upcoming mission as we worked on the submersible, getting it ready for the tough job we were all expecting. Surprisingly, even in the middle of July, the brisk North Atlantic air temperatures played havoc with SCARAB's aging electronics. We ended up installing an electric heater in the deck-mounted control van to keep the surface electronics warm and ready. Every day, as we toiled in the cold on deck, precisely at 11:00 a.m. we heard the dual sonic booms from the Concorde as it sped across the Atlantic carrying the rich and famous.

Delayed by generator problems after our arrival in Cork, we did not reach the accident scene until 6:00 a.m. on July 19, along with representatives from Boeing, the US Navy, the Canadian Air Safety Board, the Royal Canadian Mounted Police, and Air India.

The main objective of our operation was easy enough: recover aircraft wreckage that could be used to determine the cause of the accident. The technical and operational problems we would face, however, were overwhelming. First, the airliner's wreckage was strewn over an area 5 nautical miles long and about 3 nautical miles wide, and there appeared to be little logic as to the distribution of the tons of debris littering the ocean floor. Consequently, even though a terrorist bomb was suspected of causing the accident, we did not know where on the plane the bomb would have been located, and therefore where we would look for the piece of metal showing the proof. Second, the water depth in the area was over 6,200 feet, just past SCARAB's maximum operating depth of 6,000 feet. Hauling up wreckage of any size from such a depth would not be easy. A third consideration was the weather. There was considerable political pressure on the Canadian government to come up with some answers to the disaster. But the North Atlantic was not the best place in the world to play around by starting a major salvage operation in July that would easily last into the winter.

Given these conditions, our operational program was to use the SCARAB vehicle to locate all targets (as much as it was feasible) within the 15-square-mile crash site using SCARAB's onboard scanning sonar, a Benthos 35 mm underwater still camera and video cameras to document what was found, and later a specially constructed wreckage basket and lift line to haul sections of debris up to the surface.

Any human remains sighted were to be left untouched, inasmuch as this was possible to do and still accomplish our mission objectives. Nevertheless, I soberly noted that the *Cabot* was carrying over one hundred body bags.

The first thing that happened when we arrived at the crash site was . . . nothing. We spent several frustrating days staring at the rough seas waiting for a break in the weather. While our first dive attempted was aborted at a depth of only 400 feet (due to a total loss of video and a water leak), it was soon followed by the first successful descent when SCARAB reached the bottom at 4:17 on the morning of July 24.

It was boring. We all wanted to dive right into the middle of the aircraft wreckage but had been ordered to start well to the west of the main area in an effort to locate the first items to fall off of the airliner. All we saw was lots of mud, a ship's ventilator, some 55-gallon oil drums, and an old WWII torpedo. The bottom of the ocean at 6,200 feet was stark and looked somewhat like a moonscape as SCARAB's powerful lights cut a swath through the blackness.

The following narrative is a first-person account of one typical day on Air India salvage operations.

November 1985, the North Atlantic Ocean

I roll over in my bunk, trying to put off the inevitable. Unfortunately, as usual, there is a knock at the door. "Right sir, it's time!" It is 11:30 p.m. Slowly I move upright, collecting myself and becoming fully awake. Yes, I am still here.

Moving through the ritual that has over the past five months become habit, I shower, shave, and pull on my orange coveralls. I remember when I got these coveralls. It was in the North Sea during a platform inspection contract for a Norwegian diving company. They used to be clean. Now they are faded from a multitude of washings in hot water, ripped in several places, and stained with hydraulic fluid, coffee, and sea water.

Why am I here? The money isn't all that good, the work can be extremely boring at times, I'm isolated from the rest of the world. But sometimes . . . I get to see something that most people don't see throughout their whole lifetime. That's why I'm here.

Stopping by the pantry to get a cup of coffee, I glance into the officer's lounge to see what film is being shown. Nothing new—just a movie I have seen many times. New movies are a rare luxury on the *John Cabot*. (The most popular film at that time was *The Terminator*. After a few months, we could all shout out the dialogue from memory as we watched.) The food, I reason, is usually decent for about two weeks after leaving port. But soon all the fresh vegetables are gone, and the overall quality deteriorates rapidly after that. At the end of our cruise, we don't even know what type of meat we are eating.

Next, on my way to the foredeck, I stop by the test room where our Boeing, Canadian Air Safety Board, and Royal Canadian Mounted Police representatives have been watching our progress during the operation. There, I review the planned work for our upcoming dive and pick up photographs of the wreckage we will be recovering. Although this room is normally used to monitor telephone cable repair operations, it is crammed with videotape recorders, TV monitors, communications equipment, and a large chart of the crash site.

Our Air India representative, known as D. G., has recently taken to cooking curry in the test room, which has a pungent odor. The copilot of the jetliner was one of his best friends, and there was talk that he carried on board a small package just before flight. D. G. hopes that what we find does not turn out to convict his dead friend. "I don't think he would have done such a thing, but people change . . ."

Carefully, I climb up the almost vertical ladder from the cable deck outside to the darkened SCARAB launch area. The scene is a myriad of lights, activity, and people. On the left, a ship's crew member is fastening the vehicle recovery line to the craft's umbilical with masking tape. One of my fellow technicians stands in front of the SCARAB submersible doing predive checks. The underwater lights give him a strange stroboscopic appearance as they flash on and off, verifying their operation. It is cold and wet on deck with occasional spray flying over the port railing. But work continues as though the conditions do not even exist.

I slide down to the SCARAB operations van, almost slipping on a small puddle of hydraulic fluid mixed with seawater (a lethal combination for anyone working on deck), and pull open the door to check the predive progress. Inside, Gibson is diligently going down the predive checklist, running thrusters, checking the sonar, and mouthing out the systems checks. The interior of our office has long since been transformed into a small apartment about the size of a large closet. Empty coffee cups, fragments of food, Coke cans, magazines, clothing, copies of wreckage photographs, and one snoring SCARAB pilot litter the inside of the van. The wreckage photos scattered over the SCARAB control console are marked with large red Xs signifying the proper attachment points for our recovery grippers.

Back at the submarine, I give the vehicle a customary look-over in the chance that something has been missed during the predive checks. It sits on deck hunched over like a giant insect waiting to fly. Two huge orange electronics bottles on the top are infested with numerous underwater connectors that lead into the guts of the vehicle, signified by a maze of piping, cabling, video cameras, underwater lights, propulsion units . . . all haphazardly hung on a white framework. Two imposing manipulators stick out of the front of SCARAB, making it appear like a metallic praying mantis. From the submarine's rear end dangles a thick black cable through which passes SCARAB's lifeblood—2,300 volts. The vehicle is inactive right now, but with the addition of human operators on the surface it will come to life, lumbering over the seafloor like an underwater dump truck.

SCARAB is not sophisticated. What other vehicles accomplish with finesse, SCARAB does so with brute force. It is also made up of old technology. But so far, it has been extremely reliable, working at times for as long as five days on the bottom. It's almost getting to the point where we hope it will break so we can have something different to do.

I am on the CCGS *John Cabot*, 150 miles off the Kerry coast of Ireland. Strapped to the deck is SCARAB, a 3-ton remotely operated underwater vehicle. We have been on this side of the Atlantic for over five months. I left home on June 5 for a two-week cable repair operation. It is now November, and although I am only thirty-four years old, I feel like I am a hundred. I am tired from working twelve-hour shifts, seven days a week, for a month at a time. Our job is to locate, identify, and recover wreckage from a crashed airliner 6,250 feet below the surface and hopefully help Air India figure out why it crashed. Here on the ship, there is life. However, a little over a mile below in the black cold of the deep ocean is death. Death for an Air India 747 and its 329 unlucky occupants. The seafloor is peppered and strewn with a graveyard of objects: chunks of fuselage, five fanjet engines, landing gear, suitcases, galley equipment, clothing . . . and people. There have been tremendous forces at work, forces that do not discriminate between the dismemberment of an airliner and the passengers inside it.

Fortunately we, the pilots of the SCARAB vehicle, are isolated from the carnage underneath us both electronically and by distance. The passengers have long since been transformed into skeletal remains not indicative of the people they once were, or in some cases, ghostly inanimate forms with no identity. However, occasionally, the reality of what we are doing shoots through us like an electric shock . . . by the task of prying a piece of women's clothing out of the submersible's propellers, or the examination of a fragmented suitcase we have recovered, complete with name tag.

A few minutes after midnight, our crews change shifts, one happy to be finished with work for the day, the other not looking forward to a long, boring twelve hours of listening to the almost musical sweep of the sonar, plowing the vehicle through mud and between shards of ripped aluminum, and stretching beneath a dripping cable while coaxing its 7,000-foot-length onto a huge reel.

"Where's Olsen?" I ask the bridge.

"He's sick, so you'll have to work shorthanded."

Olsen has contracted a very bad cough and fever, unquestionably related to the hours all of us have spent outside in the wet cold. *That's great*, I think. *Now we have to work the shift with no break.* I feel selfish and guilty even thinking like this. Olsen needs rest. We will manage. I stuff the earpiece of the Bell Telephone StarSet into my head and lean back in the SCARAB pilot's seat, awaiting the conclusion of launch preparations.

The earpiece crackles, "Op hut, bridge."

"Roger," is my one-word reply.

"Curt, we're going to have another look at target 137 and try to find the right-hand side cargo door. Should be on a heading of 280 after you reach bottom. You all set?"

"Yeah. The crew has just gotten on deck, and we should be launching in a few minutes."

Fowler, my sonar operator, almost lies on top of the submarine as the hydraulic crane is swung over the top while spitting out the lift line. Mason, the first officer of the *Cabot*, uses hand signals until the boom end is positioned. A wave comes from the top of the vehicle. "Alright, release!" I flip on SCARAB's 35 horsepower hydraulic pump. *Whiiiiinnnnneeeee.* The high-pitched sound reverberates through my headset. I raise the red switch cover and thumb the release. Fowler stuffs the end of the lift line into the latch on the vehicle top. He motions.

Mason yells, "Alright, reset!" I jab the switch the other way and let the protective cover fall. "Secure hydraulics!" Fowler tugs at the latch, making sure the 6,000-pound vehicle will not fall as it is lifted and hops to the deck. The deck crew yanks at the manila steadying lines as the vehicle lurches up.

"Hoist up!" *Clang!* It hits the bottom of the crane latch. The crane hisses as it pulls SCARAB up and over the port rail.

"Hoist up, open latches!" The vehicle swings slightly and begins rotating as it heads for the water 20 feet below.

A call from the bridge: "Hydraulics on! Thrusters on!" SCARAB wallows in the sea as nine propellers churn the water. "Release! Thrust ahead!"

"Roger, thrusting ahead," I say as I push the joystick forward on the ultimate five-million-dollar video game.

"Pay out!" I twirl a knob on the upper right of the pilot's console. The cable counter-clicks.

"Stop!" A crew member attaches a cable float.

"Pay out!" The counter clicks.

"Stop!"

The laborious procedure goes on for about twenty-five minutes, until we have enough of SCARAB's cable floating on the surface to dive the vehicle.

Finally, we are in a good position. By now, not much can be seen of the vehicle from the ship except an unearthly glow from its underwater lights and a flashing strobe in the distance. Lawson, the project manager, calls down from his windswept post on the port wing bridge, "Hold 280 and dive!"

"Roger, diving!" I kick the vehicle around against the tug of the umbilical and move the left-hand stick forward. Slowly, the thrust meters tick ahead until they are pegged at the spot marked "Down." Some 30 or so yards from the ship, out of my sight, two plumes of water erupt upward from SCARAB's vertical thrusters and force it below the surface. Now begins the long descent to the bottom.

As the vehicle sinks, the glowing digital depth meter follows its progress: 20 meters . . . 90 meters . . . 150 meters . . . Then, as though protesting, SCARAB's starboard thruster reverses itself, definitely not to my liking.

Damn it! I take my right fist and whack the right edge of the console. The meter switches back to ahead and all is well again. I turn off the lights inside the van, and with my head propped up by my left hand, fall into a semiconscious state.

While I remain in the control van in relative comfort, Fowler, out in the weather on deck, monitors the SCARAB electrical umbilical as it unspools itself from our huge winch. Getting over 7,000 feet of cable off the winch and back on with some degree of organization is not easy. It's rather like unraveling a giant ball of string and then trying to roll it back up again. It is not a particularly safe job, either, working in such intimacy with heavy equipment.

Ninety minutes have passed. The SCARAB console is lit up by red, green, and amber lights, all signifying conditions I note with questionable interest. The depth indicator now reads 1,900 meters, not quite a nautical mile below the surface. My mind starts to wander . . . I'm falling asleep. *Wake up!* I tell myself. The altimeter, which has up to now been wavering between assorted meaningless numbers, finally locks on, telling me that the bottom is coming up and I'd better get busy. "Thirty meters off the bottom!" no longer is it necessary to yell. Normally, I would not be alone, but with Olsen out I have a lot to do. I stabilize the vehicle on the proper heading and give it some ahead thrust.

Working from memory, my fingers glide over the switches and knobs, turning on the rest of the lights, positioning the three video cameras, turning on the sonar, and rotating the range switch to the spot marked 100 meters. The bottom comes up. I ease the beast ahead hoping to make a soft landing . . . *Oops!* I can almost hear a *thunk* as SCARAB kicks up a small dust cloud while bouncing off the seafloor.

Fowler comes into the hut and takes over the sonar position. Slipping on the headphones, he pulls out the sonar chassis for a more comfortable operating position and closes his eyes. I am holding heading reasonably well as the vehicle skims quietly over the bottom.

The bridge calls. "Target should be bearing 020 relative at a range of about 60 meters. Anything yet, Fowler?"

"Nope."

A few minutes pass. The amber line of the sonar rotates until . . . *Riiinnnggg.*

"Got it. About 40 meters out, come 15 degrees right."

I turn slightly and reply, "Roger, 15 degrees."

The bright splotch on the display slowly moves toward the center while the speaker emits a variety of sounds.

"Left 5."

I give the bow thruster a burst. Fowler gives me a dirty look because the prop noise interferes with the operation of the sonar.

"OK, dead ahead . . . Right . . . Left slightly." We make a good team. "Ahead . . . OK, target about 10 meters out. Should be able to see it."

A bright panel embellished with dark paint comes into view. It is torn down the center. On the left side I make out the word "Your." On the right side I can see "palace in the sky." We have arrived.

Moving over to the right of the panel, I make out a huge main landing gear assembly. The tires have been collapsed by the water pressure.

"OK . . . Curt, what we want to do is to follow the front of this section up to the top of the wing and down the other side and see if the fuselage extends to the right-hand forward cargo door area."

Acknowledging the command from the bridge, I try to think about where everything is in relation to the vehicle before I move. Operating a remote vehicle requires the ability to mentally transport yourself to the ocean floor, using all the information and feedback you can get from the operator's console. This particular target, 137, is the largest one we have found so far. It consists of the center section of the aircraft and is lying on its right side with the left wing ripped off at the number two engine. The topmost wing end towers nearly 70 feet off the bottom and terminates in an almost flowered arrangement of wing stringers loosely dangling from the end. The lower wing skin panels have been peeled off, exposing the interior wing construction. The fuselage section is open at both ends, with some of its interior fixtures spilling out of the one end we have seen.

Cautiously, I ease off the vertical thrust and wait for SCARAB to rise off the bottom.

"Bridge, op hut. I'm starting up the leading edge now."

The vehicle slowly glides up the landing gear strut as I try to keep everything in sight. What worries me most is what I can't see, such as wreckage out of the view of the video cameras becoming snagged on the submarine or its long umbilical cable.

As I drift up the wing edge, all that appears to be left of it is the interior box structure. As I near the top, bits of debris begin to drift down past the camera, telling me that SCARAB is coming up underneath some wreckage.

"Little bit of a problem here," I report. "Let me back off a bit and see what's around."

I back SCARAB up until I can just see the wing. Noting my present heading, I swing the vehicle left and right and, sure enough, I have come up alongside a large piece of wing skin. Moving the vehicle around like this is dangerous, because if you lose sight of an object at mid-water, you are suddenly completely lost, not knowing which way to turn. The tug of the vehicle's umbilical and subsea currents can then pull the vehicle right into something you very much want to stay away from. As a result, things can rapidly start to snowball before you have a chance to use the sonar. This is not fun.

"Alright, I think I'm clear."

I continue up to the top and do a pirouette around the end before backing down the trailing edge. Controlling SCARAB in this manner is difficult, mostly because the submarine is over ten years old and does not have the latest auto-control functions. The only auto-function the robot has is an altitude feature that maintains the submarine at a selected height above the seafloor. To do a rotation around an object requires that I first move the rear of SCARAB in one direction, then use the bow thruster to compensate for the sub's turning moment. Simultaneously, I must apply axial thrust to keep the craft close to the object while down current, then away from the object on the up current side. In addition, all these functions have to continuously compensate for the vehicle being pulled by its umbilical via the ship. To do all this at the same time requires intense concentration, more hands than I have, and some degree of hand–eye coordination. Driving SCARAB at these depths can be easy or nerve-racking. Sometimes the vehicle is stable and receptive to every command, but at other times it must be tamed like a horse and forced to cooperate. Tonight, it is being reasonable.

Finally, I can see the open end of the fuselage. Rows of passenger windows, some intact, line what is now the top of the aircraft. Fowler can do nothing to help me except change the lighting on the vehicle as needed. Bright aluminum can reflect the light from the vehicle to the extent that you are, in essence, blind.

On the darkened bridge of the *John Cabot*, the helmsman stares at the surface navigation display, trying to keep the ship as near to directly above SCARAB as is possible. The first mate and the navigator make bets on how many times the helmsman can get the ship at zero offset—exactly above the submarine.

Fowler nudges me. "I'm going to go out to check the generators. Want a cup of coffee?"

"Yeah," I say, "and see if there is anything in the pantry to make a sandwich."

Working on the midnight to noon shift, I don't usually get any food until breakfast. An empty stomach helps me stay awake. Wrapped in his bright orange exposure suit, and with the aid of a flashlight, Fowler makes his way to the stern of the *Cabot* to check the oil on our two 100-kilowatt generators. They have been working so hard that the exhaust manifolds are glowing red-hot in the darkness.

By this point in the dive, my back has begun to hurt from sitting erect in the pilot's seat, which is not adjustable in any real sense. I try to move my legs up, but they hit the bottom of the control console, and moving them to the right interferes with a large gyrocompass mounted to the floor. The inside of my right ear is also starting to ache from having the StarSet ear receptacle stuffed inside it for several hours. I am beginning to have trouble focusing on the 9-inch video monitor that is less than 2 feet away. I am very uncomfortable, but I must forget all these things and concentrate.

I maneuver SCARAB over the broken-off end of the 747 and start moving closer to the bottom.

"Curt, see what there is inside the fuselage."

"Roger."

Slowly, I guide the sub in between sections of skin structure and partially inside the aircraft. I must keep the vehicle exactly centered, or it might get tangled up in the nest of cables, hydraulic lines, and debris scattered everywhere. The vehicle's lights illuminate the aircraft interior, revealing total and complete destruction. All of the internal fixtures had broken loose and are piled up inside. Very little is recognizable. It looks as though someone detached everything inside and used a giant hand to shake the airplane. A featureless humanoid shape sticks out of the pile of debris.

Backing out of the aircraft, I glide SCARAB over the top of the fuselage and along the passenger windows. I stop for a moment and park the vehicle on the side of the airplane. It is tenaciously balanced between downward thrust, currents, and the tug of the umbilical. Particles that look like snow drift by the video camera as an indication of the current rippling past the vehicle. The submarine starts to rock and slide across the fuselage. I fight it and get it back into position. The other wing of the 747 is folded over on itself and lying against the fuselage.

Outside, the wind and seas have begun to pick up. The sub is effectively isolated from the surface conditions as long as the *John Cabot* can hold position against the wind. Recovering the sub, however, is a different story. The bow of the ship surfs over a 15-foot wave and crashes into the ocean. Inside the control van, I almost fall out of my seat as the impact and vibration travels down the length of the ship, rattling dishes and waking people up. I can hear the spray raining down on the outside of the control van.

"Jesus Christ!" I complain to the bridge. "I can't drive this thing with one hand, you know!" I must grasp the edge of the console to stay in my seat. We need seat belts.

Sliding off the aircraft, I guide SCARAB to the bottom, where we think the cargo door is located. Attached to some bits of wreckage is the target we have been looking for . . . the forward right side door.

Lawson calls from the bridge, "Get yourself positioned and see what you can do to attach a gripper to the door."

After a few minutes of fumbling around and stirring up the mud, the vehicle's manipulators are within range of the door. Now that we are stopped, I can relax a bit while Fowler flips various toggle switches to position the gripper on the piece of wreckage. The gripper fitted to the vehicle's right manipulator looks like a giant ice tong and is designed to puncture aircraft structure. After attaching to the door, the gripper will be mechanically released from the manipulator end. At that point, the target will be connected to the surface by the lift line and a strain relief on the side of SCARAB. The lift line, which is held to the side of the submarine, is made of nylon-wrapped Kevlar and has a breaking strain of over 100,000 pounds.

After about 15 minutes of painstaking work, Fowler has the gripper attached to the door.

"Bridge, op hut. We're all set here for a release." Fowler turns the interlock key, raises the red protective cover, and hits the switch. The gripper falls off and raises a small dust cloud. The manipulator is stowed, and I turn the vehicle slightly to keep the lift line clear of the arm.

I call the bridge. "OK, you can start taking up the line." The helmsman of the *Cabot* maneuvers the ship until it is directly on top of the submarine. Simultaneously, the crew down on the cable deck fires up powerful cable winches to start taking up on the line. I must hold position until I see an indication that all the slack in the line is taken up, to reduce the chance of the line fouling in our own umbilical. If done incorrectly, the procedure can result in both the lift line and vehicle umbilical becoming tangled together. The least that can happen is that much time will be wasted clearing the two lines. The worst that can happen is that the vehicle umbilical will be severed and the submarine lost in deep water.

Finally, I see the submarine rocking as it is being pulled by the *Cabot*. The movement starts to become violent.

"Looks like you've got all the slack out. Releasing the lift line," I say. Suddenly free of its connection to the lift line, SCARAB smooths out as I begin driving across the bottom away from the area. "OK . . . Bridge, give me a recovery heading." (The recovery heading is the course I must drive that will take our expensive toy away from both the ship and the lift line.)

"Drive 270 degrees relative."

"Roger," I reply.

Because we are starting our ascent, Fowler must go out on deck to help feed our umbilical back onto the huge cable reel. It is a tedious procedure that none of us like to do because, first of all, it is outside, it is wet, and it is dangerous. Fowler leans over the front of the winch and under the umbilical while he talks to me over the intercom. "Take in." A minute passes. "Stop." Fowler stuffs in a piece of rope to help the cable lie properly. "Take in." Pressurized streams of seawater spurt out of the umbilical and onto Fowler's head. "Stop." He sticks in another length of rope.

Because of the many times we have dragged our umbilical through wreckage 6,000 feet from the vehicle end, the cable's outer jacket has sustained numerous cuts. The cable has become pressurized by the seawater at depth. The net result is that water spews out of the umbilical as it is being fed onto the cable reel. Unfortunately, Fowler must stand directly under the cable to see how it is lying on the reel. The same cable is passing over 2,000 volts and the deck is very wet.

At the same time, the *Cabot* is taking in the recovery line and pulling our find to the surface. The process will take over five hours. D. G., the Air India representative, and the Royal Canadian Mounted Police explosives expert are both anxious to see this section of wreckage because it might have been very close to the explosion.

Now that we're clear of the seafloor, I can just concentrate on keeping the vehicle on a general heading, away from the ship and up to the surface.

"Bridge, depth 600 meters, holding 280 degrees," I announce.

Finally, after two hours, the SCARAB's lights can be seen near the surface off the side of the port rail as an eerie yellowish-green glow. The deck recovery crew has been summoned and is filing up the ladder to the foredeck as they strap on their life preservers. Mason takes the deck headset from Fowler and begins communicating with Lawson, who has now taken position for recovery on the port wing bridge. The cargo door is still slowly being lifted to the surface.

The bow of the *Cabot* begins to turn into the wind on its bow thruster for recovery. It is starting to get lighter as dawn approaches. Lawson calls down from the bridge, "Curt turn to 100 degrees and start driving in." The normal procedure for recovery is for me to get the vehicle to the surface and then to be directed by the bridge until it is within about 50 feet of the side of the ship. Then I will switch control of the vehicle over to Lawson, who will visually make the final approach to a position underneath the recovery crane. Unfortunately, this recovery will not be easy because the seas and wind have picked up since we began our ascent over two hours ago.

"OK, Curt, I'm taking control," Lawson calls from the bridge. I reach forward and flip a switch from local to remote. He has it.

Lawson maneuvers the craft under the crane, but the seas are causing problems. As a result, he commands the crane operator to switch to "fish pull" mode. This will keep even tension on the lift wire and slowly pull the submersible into proper position before it is lifted clear of the seas. Finally, all is good.

"Hoist up!" calls Mason. SCARAB lurches clear of the water.

"Lights off!" Lawson yells. I hit four switches at once. "Hydraulics off!" The sub's hydraulic motor goes silent. "Power off!" I turn the switches marked

"Power" and "MUX" to the off positions. The orange sub is swinging bad but still headed up to a sure impact with the base of the crane latch. *Bang!* The vehicle is being steadied by the action of four shock absorbers. Now the locks are in, and the crane begins to lift the submarine and swing it over the rail. The deck crew spreads out in a circle to help guide it into a large aluminum tray bolted to the deck.

Eventually, the sub is on deck and then immediately being tied down with nylon straps. It's over. Lawson, perched up high above us, yells down, "Great job, guys!" We yell back obscenities and give him the finger.

During the post-dive inspection, I find new gashes in the sub's skids, two leaking hydraulic fittings, a dead ratfish in the aft lateral thruster (the fish has been extruded through the protective screen), a man's shirt in the port axial prop, and a loose clamp on the starboard electronics bottle end bell. I also remove a small net bag from the rear of the vehicle. Inside are about a dozen Styrofoam coffee cups, all reduced to about a quarter of their normal size by the water pressure on the bottom. Later they all are inscribed with various messages, most related to the job: Air India Salvage Ops—1985. Souvenirs.

Three hours after the recovery, the forward cargo door is hauled out of the water. The weather has worsened, and the giant bow of the *Cabot* is beginning to carve up and down in the seas. Spray is being whipped off the wave crests by the wind. By now, most of our crew is huddled near the bow—our job is finished. The door is finally clear of the water and stuck at the forward bow sheaves. It's too big to fit over the bow. The aluminum angle section our gripper is clamped to begins to rip. Subconsciously, I try to reach out and grab the door. *Twang!* The door crashes into the water and sinks. All that's left is the gripper and a fragmented piece of metal.

We spent the next two days looking for the lost door but were unable to relocate it. Our sonar was only partially operational, so we never really had a chance of finding it. By the time we had everything working, mission priorities had changed and we were ordered to look for other things. I still wonder what secrets that door could have told us had it had the opportunity . . .

———

Salvage operations on the Air India jetliner continued until December 1985, at which point the weather in the North Atlantic precluded any more work in the area. The last dive we made was to recover a wreckage basket previously deployed on the bottom that was full of small bits of debris. But by the time the basket was hauled to the surface, most of the wreckage pieces had been washed out by wave action. Approximately thirty sections of the aircraft were recovered overall, including sheets of fuselage skin, cargo, and passenger area flooring (complete with rows of seats, fortunately unoccupied), suitcases, the copilot's seat, a section of the aft pressure bulkhead, and the body of one headless man, still strapped to his business-class seat.

The official conclusion as to the cause of the disaster was that a bomb explosion had destroyed the aircraft in midair. A Cork, Ireland, newspaper reported, "Quoting sources close to India's official inquiry into the disaster, the news agency [the Press Trust of India] said the explosion lasted 30 seconds and that the printouts [from the flight data recorders] suggested the airliner began breaking up seconds later." Responsibility for the disaster was claimed by Sikh terrorists. Unfortunately, the exact details of the evidence supporting the bomb theory were never made public.

Throughout the operation a few interesting coincidences occurred. During the course of our many travels through the crash site, there was one dive where SCARAB was used to pick up suitcases, the logic being that one of them might have been near the explosion point and therefore contain fragments of a pyrotechnic device. On another dive, several pieces of baggage were picked up and then discarded because the stitching had rotted and they were simply falling apart (usually exposing books, clothing, toilet articles, etc.). However, on this dive, one suitcase was brought to the surface by the manipulator. Upon examination of the name tag, it was discovered that it belonged to a woman whose street address was John Cabot Way (the same name as our ship).

The second involved an Air India pilot who changed his schedule at the last minute. As reported by the *Limerick Leader*:

A Senior Pilot with Air India, whose flying career was saved by a miracle at a Novena session in Limerick's Mount St. Alphonsus, now has to thank the Solemn Novena a second time—for the gift of life itself. Captain M.P. Srivastava, based at Santa Cruz airport in Bombay, visited the Limerick church last year when threatened with "grounding" because of a serious heart condition. Astonishingly though, he was soon flying around the world again, and a few weeks ago he benefitted from the miraculous effects of the famous

Novena once more when he visited the church instead of piloting the doomed trans-Atlantic flight which crashed off the Kerry coast, as he was scheduled to do. A local priest who knows the captain commented that, "he has marvelous faith for a non-Christian."

A final coincidence in the Air India disaster is related to the one cargo door we attempted to recover and lost. This door was identified by the onboard Boeing representative as coming from the right side of the aircraft, forward. Out of all the cargo doors we found, it turned out to be the only one that was ripped in half. Three years after the loss of the Air India jet, a United Airlines 747 suffered a catastrophic failure of one of its cargo doors that succeeded in ripping out a huge section of fuselage, along with several passengers. The door that failed was also located on the right side of the aircraft, forward. I think it was indeed a bomb, one that detonated near the one cargo door that was ripped in half.

The *John Cabot* departed Ireland in late December to begin a harrowing fourteen-day trip back to St. John's, Newfoundland, through two full gales. Fortunately, the SCARAB system was welded to the deck, so it came though the trip relatively unscathed. However, the *Cabot* was not so lucky. By the time she limped into port, she was down to one engine, had lost both radars and her radio mast, and had suffered a fracture in a port ballast tank. Some of the interior decking had also been shored up with wooden timbers because of buckling of the plates during the transit. At one point during the trip, the captain of the *Cabot*, P. J. Chafe, spent over thirty-six straight hours on the bridge. Lawson, who rode the ship across during the storm, told me that he slept on the floor near his bunk wearing his exposure suit in case the ship went down.

Instead of riding the ship home across the ocean, I flew back because by that time my body was failing after so many twelve-hour shifts. Plus, I had a bad cold and simply could not work anymore. My flight back to Maryland to pick up my paychecks was in a similar Boeing 747 jetliner. I was sitting right about the same location where the *Kanishka* had broken in half.

I stayed in the condo of high school friend Bill Wood in Ballston, Virginia, where I set up a meeting with the National Air and Space Museum to pitch the idea of finding and recovering Gus Grissom's *Liberty Bell 7* Mercury spacecraft. Bill's father, Col. William "Woody" Wood, had flown with Grissom at

Presque Isle, Maine, before being sent to Korea. I had a good time hearing the stories. After meeting with NASM, I headed to Colorado.

After arriving in Colorado, where I was staying with Dad and wife number three, I tried to collect my thoughts about the whole thing. Incredibly, after about two days at home, I was asked to go out on another job—a platform inspection. For the first time in my career, I turned down a job. I desperately needed a break from working on a ship and decided to head out to Southern California to spend some time with my brother and relax. It was an enjoyable respite, checking out new rock bands at the Troubadour, hanging out on the Sunset Strip, sitting in with my Fender Precision Bass at the Central, and reconnecting with some high school friends who were living there at the time.

I was in a deep, alcohol-induced sleep when I felt someone shaking me. The date was January 28, 1986. Though still half asleep, I remember my brother, Chris, saying, "Get up. The space shuttle just blew up." *Shit.*

8

1986: *Challenger*

Florida—January 28, 1986. The launch of the twenty-fifth space shuttle mission. I was in California visiting my brother when *Challenger* was lost. Like other Americans, I watched with a heavy heart as another American spacecraft was lost at sea. Unlike *Liberty Bell 7*, this one did not remain intact, and its location would be marked by tangled debris drifting in the Gulf Stream currents. Tons of wreckage peppered the seafloor, much like the Air India jetliner, and once again, it would be up to people like me to help find the *one* piece of wreckage that provides conclusive proof of the cause of the disaster. The salvage of *Challenger* was the largest search and recovery operation in history and required the use of a mixture of underwater technologies: side-scan sonars to map the debris field, manned submersibles to identify wreckage, and divers and remote vehicles to recover the evidence.

I watched the videotape of the accident over and over as the commentator described rescue efforts. It was painfully clear that there would not be any survivors.

Two weeks later I received a phone call from Eastport International (formerly Ocean Search, the company I had worked for in Ireland on the Air India job). They wanted me to go to Florida and pilot the Gemini underwater vehicle and help pick up solid rocket booster wreckage for the US Navy's Supervisor of

Salvage and Diving. I arrived in Cocoa Beach on a brisk February morning in 1986 to help mobilize our massive submersible system for the job. It was going to be a difficult operation because we would be working right in the middle of the Gulf Stream, where 3- to 4-knot currents were commonplace. While I was still burned out from the Air India job, I had been on the beach for six weeks and was as ready as I could be.

The plan so far was for the Gemini to work off the DSV *Stena Workhorse*, a 313-foot-long heavy salvage vessel manned by a Swedish and Finnish crew. The *Stena* had considerable deck space for recovering wreckage and two huge 50-ton cranes located near the stern; it would be ideal for the job. The Gemini was a 5,000-foot-capable remote controlled underwater vehicle used only once before to salvage a crashed helicopter in the Pacific. While it was developed by the same manufacturer as SCARAB, Ametek-Straza, it was configured differently in that it was deployed underwater from a top hat cable management system. In other words, the Gemini and its submerged launching system were deployed together. The Gemini would be maneuvered underwater only when it had been detached from its launcher. Its powerful 80-horsepower propulsion system would help it work in the strong currents we expected to encounter.

We were not the only group working on the *Challenger* salvage. The US Navy had dedicated several salvage vessels of the ARS (auxiliary rescue and salvage ship) class, such as the USS *Opportune* (ARS-41), to supply divers and act as platforms for the operation of smaller remote vehicle systems. In addition, the Harbor Branch Foundation supplied both of their Johnson Sea Link manned submersibles to help recover small wreckage parts and identify objects located with side-scan sonar. They would prove invaluable in this role. Morton Thiokol's booster recovery ships, such as the MV *Liberty Star* and *Freedom Star*, assisted the search by acting as platforms for towed side-scan sonars. The MV *Independence* was the work platform for the Navy's Deep Drone ROV (remotely operated vehicle).

The task confronting the Navy was overwhelming: search an area encompassing about 470 square nautical miles and identify all targets as being *Challenger* or non-*Challenger*, inspect and categorize the targets, then recover all wreckage that might bear evidence of the disaster. Unfortunately, the location where *Challenger* went down was heavily traveled by ship and air traffic and drug smugglers, and it had been the repository of a large percentage of NASA's launch failures. There was a lot of space junk littering the seafloor.

The Navy set up their priorities as follows. First, they simply had to find the smoking gun. While it was strongly suspected that a segment of *Challenger's* right-hand booster had failed, NASA had to be sure. In the tons of debris stuck on the bottom, we had to find that *one* piece. Second, for humanitarian purposes, the Navy had to recover the astronauts' remains. Based on my Air India experience, I knew there would not be much left to recover. *Challenger's* crew compartment had struck the ocean at over 200 miles per hour and broken up into several pieces. Third, we had to find and recover the tracking and data relay satellite located in the shuttle's cargo bay. If it was not found, the government would have to spend millions of dollars to change satellite communication codes so the Soviets could not find the TDRS and subsequently monitor our military communications.

Getting the Navy's Deep Drone vehicle to Florida was a daunting task because it was in the middle of a major refit when *Challenger* went down. As John Finke, a Deep Drone team member at the time, recalls:

> We were in the shop in Upper Marlboro, Maryland, and Craig Mullen, the president of the company, told us to turn on the television as there was an incident at Cape Canaveral. We got a television and saw that the *Challenger* had blown up. At the same time, we were in the lab with the US Navy's Deep Drone vehicle, and it was in the middle of a breakdown period. We had taken the electronics bottles off the frame . . . the thrusters were off the frame; it was totally in breakdown mode. . . . We all looked at each other and said, "Well, what are we going to do now?" So, we immediately at that moment started to put it back together again.

It took John and his team about five days to get Deep Drone ready, then they took it to Andrews Air Force Base with a police escort and loaded it on a C-141 cargo jet and flew it to the Air Force base at Cape Canaveral. John continues:

> While we were putting Deep Drone back together, the *Independence* had been out to the estimated crash location with another ROV and was not very successful at finding any wreckage or parts of the shuttle.

After failing to find any *Challenger* wreckage, the MV *Independence* came back into port, where the crew removed the existing equipment and installed Deep Drone on the ship. They sailed immediately after completion of mobilization. There were already search vessels out scouring the area, examining areas of the seafloor where radar data indicated that the spacecraft had impacted the water. They found the debris field to be huge because the various components of the shuttle had landed in different areas. Deep Drone's immediate tasking after sailing was simply to find the area. They headed out to the estimated location of where the two solid rocket boosters might have landed, in particular the right-hand booster. But the location was right in the middle of the Gulf Stream, and the Navy concluded that they would need a more powerful vehicle to operate in that area. Consequently, the Deep Drone crew were retasked by the Navy to go back out and try to find the main part of the orbiter because it was thought to be closer to shore in shallower waters with less current.

The USS *Preserver*, with Mobile Diving and Salvage Unit divers, was tasked with finding the remains of the crew compartment, and all other assets were told to stay away from them. Once they had picked up some wreckage and "cleared the area," the Navy sent in Deep Drone and the *Independence* to see what else was left on the bottom. As John Finke recalls:

> We were given an "exact" position by the command center for the crew compartment and told to dive there and see what we could find. . . . We got to the bottom and the CTFM [continuous transmission frequency modulation] sonar lit up with targets . . . many, many targets. Having been told that the area had been cleared by Navy divers, it was obvious [to us] that there was still a lot more there that they didn't find. We started a line search with Deep Drone and started using baskets to collect assorted debris.

The captain of the *Independence*, Mark Wood, was on the ship's fantail reading a *Time* magazine article about the shuttle disaster. One of the images was a very detailed photo of an astronaut in the cargo bay that showed a light specially molded into his space helmet. "We looked at the *Time* magazine photo and then into one of our baskets of wreckage, and sure enough, there was a light just like the one in the magazine. So, we knew we were in the right area," said Wood.

After spending about ten days playing with the Gemini vehicle and finally getting it to work, we sailed to the crash site. Most of the orbiter parts were in shallow water and about 20 miles from the Cape. But the depth where the boosters landed ranged from 600 to over 1,300 feet. Each section of booster wreckage could weigh up to 6,000 pounds. Our planned technique in recovering wreckage was a little complex. Once an object had been designated for recovery, we would use the Gemini to attach two small recovery tools and a lifting sling to the booster fragment. While we had a tool specially designed by NASA for the job, we quickly discovered it did not work and subsequently developed a new tool at sea out of a modified lifting shackle. Once the wreckage part had been prepared for recovery, a long length of Kevlar lift line would be lowered down to a point down current of the wreckage. Because of the strong surface currents, the end of the line would be weighted down with a depressor weight—a 2,000-pound length of anchor chain. The Gemini would be launched and fly at mid-water up to the deployed lift line, grab the free end, drag it back against the current, and attach it to the wreckage. Then Gemini would be recovered and the booster fragment hauled up to the surface.

Once near the surface, the *Stena*'s deck crew would attach crane lines to the part, drag it out of the water, and dump it on the *Stena*'s oily wooden deck.

One major issue was the unburned solid rocket fuel still on *Challenger*'s booster fragments. The engineers from Morton Thiokol simply didn't know what to do with the wreckage because there were no procedures for removing unburned propellant. To make matters worse, some of the Thiokol people were saying that as the wreckage cleared the surface, the rocket fuel could spontaneously ignite. Even if the booster wreckage did not ignite when it cleared the surface, simply having any significant amount of the stuff on the *Stena*'s deck would not be the safest thing in the world. The engineers I talked with soberly noted that the high temperatures created by the burning of only *one* wreckage part on the *Stena*'s deck would probably burn a hole right through our deck and sink the ship.

To test our salvage techniques, we headed out to the area where *Challenger*'s left-hand booster was located. Our logic was that we wanted to test out our techniques in 600-foot depths before we moved into 1,300 feet of water. We ended up recovering a small section of left-hand booster, probably weighing no more than 1,000 pounds. But at least everything seemed to work, for now anyway.

We began diving at the location where the right-hand booster had landed, all to find the one burn-through piece everyone wanted. It was during the next few weeks that we started having problems with the Gemini.

Contemporary unmanned submersibles are connected to the surface via an electrical umbilical. In the case of the Gemini, there were two such cables. The first ran from the *Stena* to a submerged launching system and was constructed out of a heavy steel-armored cable. Inside of this cable ran the electrical wiring needed to power and control the vehicle. Running from the top hat submerged launching system was another lighter cable made up of a strength member manufactured from Kevlar. Like the steel umbilical, this second cable contained a considerable amount of electrical wiring. The problem was that this second cable was not nearly strong enough to support the drag of the powerful Gulf Steam currents and the beating it would take as the Gemini was maneuvered on the bottom. Consequently, the wires contained within this cable broke—*a lot*. Every time these wires failed, we had to do what is called a re-termination, where the cable is cut off and rewired into the Gemini, a process that took up to ten hours to complete. On a normal remote vehicle job, it was considered a pain if you re-terminated more than once.

By the end of our operation, we had re-terminated more than thirty times. We also had used up four of these lightweight umbilicals in the process. As a result, the cable ended up being a disposable item of sorts. As the operation proceeded, the failed terminations would end up stacked on our workbench like some sort of monument to the unreliable nature of the Gemini system. One time, the cable failed in a spectacular manner that resulted in a desperate search for the vehicle as it drifted north in the strong currents.

———

It was midnight as I came down onto the *Stena*'s back deck. Barry Brown, one of our most experienced pilots, was driving the Gemini as it was being recovered. On deck, the Gemini's deafening generator droned into the night as our support personnel monitored the progress of spooling the Gemini cable onto its giant winch. Lawson, my comrade from the Air India job, stood on our elevated launching platform. His coveralls were stained with hydraulic fluid and grease as he continuously updated Barry with our slow progress in recovering the vehicle. I could see that things were not going well.

As Lawson lifted one side of his headset, I yelled to him, "How's it going? Do we still have a working vehicle?"

"Well, sort of . . . The launching system is jammed, and we are trying to bring up the Gemini on its own."

I knew what that meant. While the Gemini should normally be mated to its submerged launcher, it was still dangling at the end of its tether. Not a problem now while the Gemini was still at depth, but when we got to the surface and the currents became strong, it would be a different matter.

I peered over the starboard side of the *Stena*, hoping to get a glimpse of something, anything. I noted with concern the set of the Gemini's main umbilical. It was angled heavily toward the *Stena*'s stern and was vibrating like mad in the 4-knot surface currents. This was going to be interesting. Occasionally, the *Stena* would heave up then crash down in the heavy seas, causing the cable to slack and then snap tight with a bang.

I turned back to Lawson. "I'm going to go down to see if Barry needs any help."

Down in the Gemini control room, Barry was struggling as he fought to keep the vehicle from being swept away. Unlike the SCARAB control area, the Gemini control van was spacious and manufactured out of a modified 20-foot-long shipping container. However, despite its greater size, the van's interior was still crammed with the green Gemini control console, extra video monitors, communications equipment, spare parts, storage cabinets, and a workbench. It had been thrown together in a hurry by Eastport because of the pressing nature of the *Challenger* salvage job.

The main controller for the Gemini, a black plastic joystick populated with numerous buttons and switches, was jammed completely forward, telling me that Barry had on full ahead thrust. His knuckles were white with tension as he abruptly jammed the stick over to the left, then back to forward.

Keeping his eyes glued to the digital heading indicator, Barry murmured, "You just got to keep the thing straight into the current. If you don't, it falls right off."

The Gemini had the most thrust when driven straight into a current. In our situation, every time the currents hit it from the side, it would start to swing. If it went too far, Barry would be unable to stay close enough to the submerged launcher. We couldn't pay out more cable from the launcher because the cable drive mechanism was jammed. If the vehicle was swept down current from the launcher and the ship heaved at the wrong moment, the Gemini's light tether would snap like a piece of thread and we would lose the vehicle.

Noting that it was past midnight and Barry was off shift, I offered, "If you need any help or get tired, let me know and I'll take over."

He didn't hesitate. "It's all yours!"

I quickly sat down in front of the control console and slipped on the headset, keeping the control stick full forward as our hands traded positions. Because we were off the seafloor, all I had was a clouded view of the submerged launcher and its glowing light. Particles of matter drifted toward me on the video monitor as I fought the vehicle. The only way I could keep the Gemini close enough was to maintain the launcher in sight. If I lost it, I would have no idea of which way to drive the vehicle to get back to where I was supposed to be. Our navigation system was useless at this shallow depth.

Hitting the transmit button, I called out to the deck, "We're at 300 feet and I can tell the current is starting to pick up a bit. I'm starting to have a lot of trouble staying with the launcher."

I was losing ground. Even with full ahead thrust, I wasn't sure I could maintain control of the Gemini. Underwater, the 5,000-pound submersible fought against the powerful currents as its screaming hydraulic propulsion units shot back circular plumes of agitated seawater. The fast-moving currents whipped and vibrated its electrical cable as the Gemini rose closer to the surface.

Out on deck, Lawson could see a dim green glow of light as the Gemini's launcher rose to just underneath the churning sea.

"I can just see the launcher. How are you doing?"

"Not too good . . . I'm starting to fall down current," I replied.

"Hang on. We're almost there!"

As the Gemini rose closer to the surface, the Gulf Stream currents intensified and screamed along the side of the *Stena*. The vehicle began to drift back from the launcher and in the process tighten up its jammed electrical cable. The cable began to stretch.

I was watching the monitor when it happened. First, the MUX lock light flickered, then glowed brightly. The light was an imminent indication that all control of the Gemini had been lost. Next, the image on the console's video monitor began to shake, then quickly became obscured with horizontal lines of interference. We were losing the video. The Gemini's cable was starting to break.

"I'm losing video and I've got a MUX light!" I yelled.

The rough seas pushed the *Stena* up one last time . . . Then the video went blank and the circuit breakers tripped. Underwater, the Gemini's electrical cable was violently yanked away from the vehicle's frame as high-voltage current poured into the sea. The Gemini was lost. Frantically, I tried resetting the breakers and turning the power back on, but it was no use. The cable to the Gemini was now dumping over 2,000 volts into the ocean. The electrical breakers would not reset. Even though it was not really my fault because of the conditions, I felt responsible. In nearly ten years of driving underwater vehicles, this was the first time I had ever lost one. The thought *I lost the vehicle* churned my stomach.

I called the deck on the intercom: "Sorry—I think we lost it."

Outside on deck, Lawson saw the damaged launcher being hauled out of the water. The yellow cable once attached to the Gemini was stretched and broken, with several wires hanging out of the end. Gemini was gone.

Under the *Stena*, the Gemini rose upward, bounced off the hull of the ship, bashed into a huge propeller, and slowly came back to the surface as it drifted away from the ship to the north. Rushing out on deck, I joined the rest of the crew straining to see the Gemini. It was pitch-black on the ocean as the *Stena*'s search lights illuminated the sea in a search for the expensive vehicle. While the Gemini was fitted with an emergency strobe flasher, it was not turned on. The one time the vehicle would not come back, the flasher was off.

After about two hours of frenzied searching, we finally sighted the yellow top of the Gemini, bobbing intermittently up to the surface between the heavy seas. A short while later, the battered Gemini was safely on deck and in the process of being disassembled. We couldn't fix the vehicle at sea, so the *Stena* headed back into Port Canaveral.

Surprisingly, the Gemini wasn't too bad off. A few gashes in its top, smashed-in frame members, broken lights, some marks of hull paint . . . A few repairs and it would dive again. Two days later we were back at sea.

One of the problems we had working in the *Challenger* debris was the numerous fish that congregated near the front of the ROV, attracted by the lights. Because they obscured our vision, I came up with a "fish light." I simply mounted one of our underwater lights to the rear of the Gemini. When the school of fish was too much, I turned off the front lights and turned on the rear one. After a few minutes, the fish would become interested in the rear light and reposition themselves around the vehicle's stern. Then, I'd fire up the front light and get some work done until they swam back around. It was a constant back and forth.

We spent the next four weeks diving over and over to the 1,300-foot depths in our search for *Challenger*'s burned-through booster fragment. Again, we dumped the Gemini into the seas. Day, night, calm, rough. It didn't matter. We pulled the wreckage up, lines would snap, and again the piece would fall back to the seafloor. Again we cut off the Gemini's electrical cable and coaxed fragile wiring back into place using our battered fingers. Again and again, razor-edged wreckage would swing wildly across our deck in the night and carve deep gashes through the wood planking as it stopped with a crash. Again we filled the *Stena*'s deck with volatile booster wreckage. And finally, we found what everyone wanted to see: a 6,000-pound chunk of steel. On one edge was an opening unlike what we had seen before. It was rounded and melted, not broken and sharp. This was how *Challenger* had perished.

When we found the section of wreckage showing conclusive evidence of *Challenger*'s loss, we immediately headed into port. This time, we didn't berth at the commercial dock at Port Canaveral, but at the Trident submarine base across the channel. I had attached the lift lines to the wreckage part and struggled with low visibility conditions on the bottom and a partially functioning manipulator (partially functioning because I had banged it into the wreckage, severing a hydraulic line in the process). I was so disgusted with the dive that, unlike my normal habit, I didn't even bother to look at the wreckage we recovered—until the next morning, when one of my fellow crew members said to me, "You should have a look at that booster fragment recovered last night. You might find it interesting . . ."

During ongoing operations, the Deep Drone salvage crew on the MV *Independence* came upon what had to be *Challenger*'s crew compartment containing the remains of the astronauts. What they found was disturbing, as John Finke recalls:

The crew cabin was pieces and parts. . . . You could not say it was a structure. You couldn't tell the inside from the outside. It was destroyed. It was large wreckage, maybe 10 by 20 feet or 10 by 10 feet. . . . All just mangled metal. At one point, the Deep Drone did come upon one part of human remains, which was identified as a pelvis. At that time, we stopped all operations and brought in medical teams to recover it.

John does not remember the name of the "medical ship," but records indicate it was likely the USS *Preserver*, a salvage ship outfitted with Navy divers. They apparently recovered most of the remains of the astronauts. This work spanned several weeks. The divers worked from the USS *Preserver*, with Deep Drone used to pick up what they missed or could not find. Once they had sanitized the area, the NASA people came back in and picked up their gear.

While the Deep Drone system was scheduled to demobilize from the *Independence* and go to the Bahamas to recover torpedoes, NASA tried launching another rocket, a Delta. It did not go as expected, as John recalls:

At the tail end of this time, NASA was going to launch another rocket. We were on the deck at Cape Canaveral when the rocket took off, and there was a problem with that rocket. . . . It exploded also. It was a very bad situation with one rocket exploding after another, and personnel on other vessels were tasked to go after the most recent launch. But Deep Drone was tasked to go to the Bahamas and recover torpedoes.

Most of the Deep Drone crew from the *Independence* did not get home until six months later.

After recovering a 14,000-pound section of *Challenger*'s booster end (I drove the vehicle up inside the booster to drag a heavy chain through the opening) and finding the other half of the burned-through segment, we headed back to Florida and began pulling the Gemini system off the *Stena*. I was exhausted. I had already gotten a bad strep throat during the operation, requiring me to be carted in to shore in an incapacitated state to get well. That whole situation had been almost unbearable.

First, I was tossed off the ship into a crew boat called, of all things, the *Eliminator*. It almost eliminated me as I bounced around in the compartment below the decks, which reeked of fresh paint and diesel fumes. The boat drivers didn't seem to know any speed but full as we bounced along the sea.

When we finally arrived at the commercial port, an Eastport engineer was there with a rental car to take me to the emergency room at the county hospital. I felt weak and horrible and just wanted to go to bed. Before long, I was sitting in a wheelchair as an intern poked and prodded my throat. Suddenly, he waved a young companion over to also look deep into my mouth. "Now, *this* is what a strep throat looks like!" OK, so now I was being used as a teaching aid to some dumb guy when all I wanted to do was get some drugs and go to sleep.

Because I was too weak to walk in, my engineer friend accompanied me to the pharmacy and managed to get me some antibiotics. When he asked me whether I wanted the rental car, I declined. I could barely walk and was in no condition to drive anything anywhere. I was only on shore for two days when I was dragged back to the *Stena*.

I took a day off before getting back on another ship, the *Independence*, to drive a small remote vehicle called the ADS-620 (advanced diving system) to help identify and recover sections of the orbiter from shallow water. The job was like a vacation compared to working on the *Stena*—shallow water, good weather, and good food.

During my time on the *Independence*, we found all kinds of objects, many not related to the loss of the space shuttle. Among the items discovered were the normal "nothing found," as in there was nothing there: geology; a trawl door and cables; Pershing missiles; multiple shipwrecks, some wooden and some made from steel as long as 300 feet; an intact DC-3 aircraft; a Redstone rocket; outboard motors; fishing boats; 55-gallon oil drums; anchors; toilet bowls; and even the proverbial kitchen sink. But one object found on May 3 garnered the interest of Capt. Wood: a beautiful, intact airplane propeller. The thing was huge and in excellent condition, with minimal marine growth on it. We had a video monitor on the bridge, and it wasn't long before the captain came into our control van.

Pointing his finger at the video screen, he said, "I want that!"

"Well, you're the captain and can have it if you want it," I replied.

"We're supposed to clear the area, aren't we?"

"Yessir."

"I'm going to have my boys go down there and put on some lift bags."

It couldn't weigh too much, could it? Weren't these things made from aluminum or magnesium? It should be easy to lift. That's what we figured, anyway.

I hovered the ADS-620 ROV around the propeller, trying not to stir up the bottom too much. It was not long before a couple of scuba divers entered the area and attached a few lift bags to the prop, then filled them up with air from their regulators. The massive propeller didn't budge.

Some time passed, then a second dive team came down, doing the same thing. More lift bags. Still, the propeller remained glued to the bottom, in spite of their attempts to lift its tips.

A third team came down and used up the last of their lift bags as we watched on our video monitor. Nothing.

Finally, I called the bridge and said, "Why don't you let me use the ROV's thrusters to clear some of the silt away. Maybe that's the problem."

"OK, go for it."

So, I positioned the vehicle to one side of the propeller, put on full down thrust, and gave the area a good blast with the vertical thrusters for several minutes. After letting the current take away the silt, I maneuvered the ROV so that we could see what we had. Underneath the propeller was a massive and very heavy radial engine easily weighing over a ton.

I called the bridge again and said, "Captain, I think we have a problem."

After a long sigh, Capt. Wood replied, "Forget it. Let's get the lift bags back and carry on."

The propeller is still there.

Finally, it was all over. I told Eastport that I was worn out and wanted to go home for a while. As a last reminder of the nature of the wreckage we had recovered, I learned that a member of the salvage team had tried using a small chunk of *Challenger*'s solid rocket fuel to start up his barbecue; he melted a hole through the bottom of his grill in the process.

On another occasion, a member of the salvage crew bragged to a young woman in a bar that he had pieces of the shuttle at his hotel, and would she like to see them. Unfortunately for him, she worked for Morton Thiokol, and it was not long before several of the salvage ships were raided by the FBI, looking for souvenirs. Both ships I worked on avoided that fate.

Life on the *Stena* was not bad, though the food could have been better. Everything was always overcooked, rendering the eggs into a hardened plastic eatable. Our accommodations on the ship were 20-foot shipping containers converted into living spaces. They were better than some I had experienced on other ships. At least I had an actual bunk.

Unlike most jobs, the *Stena* had several female crew members, who were either Swedish or Finnish. The Swedes tended to look down on the Finns as low class. When we initially sailed, all the young women, whose main job was to clean the ship, sunbathed on the helideck topless, as was their custom. Of course, none of us had any problem with that at all. But one of the women was a very attractive blond who was the captain's girlfriend, so that soon came to an end.

As far as I know, all our relationships with the women on board were platonic. We worked together and partied at times when we were on the beach, but that was about as far as it went. The Swedes liked me because I told them I had just bought a Saab automobile.

Some of the other crews stayed on for several more months in a search for the TDRS once housed in *Challenger*'s cargo bay. Even after the satellite was found and recovered, the Navy had them continue to recover other wreckage in order to not raise any suspicions as to what they had been doing. The objectives set forth by the Navy had been met, even though over half of the orbiter, most of the left-hand booster, and 40 percent of the right booster remained on the bottom of the Atlantic. They will be there forever.

Two individuals stand out to me who were instrumental in salvaging the space shuttle *Challenger*. The first was Morton Uloff of Stena Line, who I believe was the first mate. He oversaw the deck anytime wreckage was being recovered. Although we were picking up razor-sharp steel wreckage covered with solid rocket fuel, no one ever got hurt. With his red coveralls and leather jacket, he controlled the deck and coaxed the sometimes wildly swinging booster fragments onto the ship.

The other was the late Capt. Charles Bartholomew, nicknamed "Black Bart." Bart oversaw the entire salvage mission and had the difficult task of coordinating and managing our salvage assets. He was one of those rare individuals who was truly larger than life. His hot tub parties at the Cocoa Beach Holiday Inn were legendary. Bart was a brilliant man who vastly improved the Navy's ability to conduct deep ocean salvage operations and should have lived much longer than his fifty years.

I met Bart several times, but the first time was in the men's room at the Holiday Inn while our crew was out drinking. I didn't know who he was. We were standing together at the urinals when he commented on my black leather jacket. "That's a nice jacket. It looks good, not like the cheap ones you normally see."

"Well thanks," I replied. "I'm down here on the *Challenger* salvage . . ." Bart sort of smirked as he told me, "I'm Adm. Bartholomew and am in charge of the operation!" Bart was a captain at the time, but I will give him being an admiral any day.

Capt. Bartholomew tragically lost his life on November 15, 1990, during a recertification dive off Panama City, Florida. My friend John recovered his body, and I attended his burial at Arlington National Cemetery. Dick Asher, an engineer I had worked with at Ocean Systems, sobbed while delivering his eulogy.

Just before I left Florida, I spent a day at the Kennedy Space Center searching through their records trying to find more clues to the location of *Liberty Bell 7*. I didn't have much luck because they had less information than the NASA History Office in Washington.

More than twenty years later, some local divers claimed that they had found *Challenger*'s right wing. We knew where it was the whole time and hadn't bothered to recover it because it was too big to fit on the *Independence* and wasn't needed for the investigation.

9

1996: TWA Flight 800

F ollowing the end of the *Challenger* salvage and over a period of several
years, I worked on a variety of operations ranging from cable work using SCARAB II and a SCORPIO 2000, Sea Plow work off Newfoundland,
an F-15 salvage near Japan, and considerable work in the Mediterranean surveying and burying an assortment of communications cables called SEA-ME-
WE 2 (Southeast Asia–Middle East–Western Europe 2). These were submarine telephone cables linking many parts of the Mediterranean and were done
from the France Telecomm ship CS *Raymond Croze*. Somewhere in there I
once again became a free agent, doing work for Margus Co., a small undersea services firm formed by two former associates in the cable maintenance business, Gus Dodeman and Bill Wall. Gus was our former client with
Transoceanic Cable Ships and Bill was a former member of the SCARAB I
Cable & Wireless team. They acquired a SCORPIO 2000 vehicle from a US
Marshals sale in Florida, required when a treasure hunter went out of business. While the vehicle was in rough shape, we generally made it work and
completed many post-lay burial operations in the Mediterranean working
out of Catania, Sicily.

We also did some work with an aging Sea Plow off the coast of Newfoundland in waters where icebergs were commonplace. I had never worked with

a Sea Plow before, and the one we were leasing was Sea Plow V, an obsolete system that eventually ended up as an ornament plopped in the grass at the entry to AT&T's cable ship base in Baltimore, Maryland. But for now, we had to get the vehicle ready to go to sea for a cable burial job using my old home during Air India, the CCGS *John Cabot*.

Compared to a work class remotely operated vehicle (ROV), Sea Plow V was a relatively simple system that used a main tow cable to drag it across the seafloor, a second cable to supply power and receive video, and a compressed gas hydraulic system to raise and lower the plow that dug the trench for the submarine cable. Unlike with newer versions of Sea Plows, the cable to be buried had to be cut off and threaded through the plow, which complicated operations.

The vehicle was made from steel and looked much like a boat, with a bow and stern. Approximately 25 to 30 feet long and weighing several tons, it was launched off the stern of the *Cabot* using two hefty davits and dragged from the stern of the ship. It was not easy to get in and out of the water. A special design factor of the plow was that the main tow cable had a weak point purposely installed so that if the plow got stuck against an obstruction, the weak point would fail and deploy a secondary length of tow cable that would string out until the ship was able to stop. It was good in theory, but not easy to employ in practice.

We easily got the plow ready for sea in AT&T's facility in Baltimore, and before long we were dragging the contraption across the bottom south of Newfoundland in an attempt to bury a section of CANTAT-3, a Canadian telephone cable.

Another unique aspect of the Sea Plow was that it was fitted with a small hydrophone that generated an audible sound that allowed us to hear what was going on with the plow as it careened across the bottom. I concluded that sometimes it is better not to know what's going on underwater, especially on a rocky sea floor. The crashing and banging sounds as the plow forced its way through the sediment were quite unnerving.

Because most towlines rotate under tension, there was also a huge swivel connecting the main towline to the plow. We typically observed this on our underwater camera, watching in disbelief as it would spin up at considerable revolutions under tension, only to spin back the other way when the tow cable relaxed. It made a lot of racket that forced us to turn down the volume control on the hydrophone because it was too disconcerting.

In the plow's control van there was little instrumentation, just a few switches and analog gauges, as well as a large paper plotting chart with assorted colored lines created by mechanical ink pens that told us various things about what was going on, such as the depth, heading, and speed.

So, we were towing this contraption in about 900 feet of water, and everything was going quite well. I could tell much about the bottom by the sounds from the hydrophone; a soft swishing sound meant nice, smooth sediment; repetitive crashing and banging sounds meant that we were dragging the plow through an area where icebergs had deposited small- to medium-sized boulders after melting. What I either missed or didn't know was that these same icebergs were large enough to create furrows in the bottom, many of them across our path. As I later discovered, some of these furrows could be quite large—large enough to stop the plow.

We were all relaxed, and I remember thinking that this was an easy job. That changed quickly. Out of the corner of my half-open eye, I saw that the bottom had suddenly gone away. The attitude of the plow took a hard nose-down angle, there was a burst of sediment on the camera, the crashing stopped, and everything went silent. Silence is not good when you're towing. We had struck an iceberg furrow; the plow had crashed into the other wall and was stuck.

The problem was that the ship was still steaming. I quickly called the bridge on the intercom. "Stop the ship! Stop the ship!" There was silence as we continued ahead. "STOP THE SHIP! WE'RE STUCK!" I screamed.

More time passed before someone got on the line. "You say something?"

A 325-foot-long cable ship does not do anything fast, especially stop. We watched in terror as the towline stretched more and more, with the swivel starting to spin like a top. It sounded like a massive drill that was groaning under tons of force. Suddenly, there was a loud bang, and everything went quiet again. Not good because our weak point had failed, and if the ship didn't stop soon, the Sea Plow would be history because we didn't have divers on board who could reach 900 feet.

Fortunately, the *Cabot* stopped in time, and we eventually lifted the plow back to the deck. Unfortunately, the test engineers told us that the cable had taken a hit and needed to be repaired.

Before long, we were once again towing the plow across the cavernous bottom. No longer were we relaxed. We tilted the scanning sonar down a bit so we might be able to see the next furrow, but we were totally on edge with whitened knuckles grasping the sides of the consoles, waiting for the next iceberg canyon. The sounds coming from the plow now terrified us even more than before as we listened to it carve a path through the boulders and rocks. But we never hit another iceberg furrow and managed to complete the job without incident. Even though the route had been previously cleared using a special cable-cutting grapple, we dredged up all kinds of things in the blade, such as nineteenth-century telegraph cables, trawling gear, and fishing nets.

Given the uncertain nature of the work, I was offered and accepted a job back at Eastport International in Lanham, Maryland. In 1988, Eastport made the decision to try to enter the oil field servicing market. I told them that based on my previous experience, it would be a mistake. The oil field service companies operating in the Gulf were well established, owned their own ships, and were well positioned for callout work. In other words, they could respond quickly when needed, and we didn't have that ability because we didn't own any ships. Eastport hired a couple of old Texans, who acted as consultants to help them out.

During the company's initial entry into this market, I was asked to operate a small Ametek Scorpi ROV to assist in a J-tube installation on a production platform in the Gulf of Mexico. The J-tube on a production platform is where the pipeline connects to the rig. The "J" in the nomenclature is because as the pipeline comes up to the side of the platform, it has about a 90-degree bend as it snakes up one of the platform's legs. There is usually a flange on the seafloor at that location, and I was supposed to inspect it to make sure the divers had assembled it properly.

We were working out of Venice, Louisiana, a small town totally beholden to the fishing and oil field industries. Based upon what I saw in 1988, it was a town that was already dying. The one movie theater had been closed for years, and many of the residents were out of work. I remember there being one bar where we drank before sailing and it was nothing much. The town was pretty much obliterated later during Hurricane Katrina in 2005.

I cannot recall the name of the aluminum boat we operated from, but it was inhabited by a series of young divers who had the ability to dive to almost 500 feet on air. This was a practice that was not often done because it was danger-ous due to the partial pressure of oxygen at that depth. But if you are young, adventurous, and acclimated to such conditions, it can be done reasonably safely. In addition, being able to dive deep on air as opposed to mixed gas is like a lot of things—some people can do it and others can't.

Given my past experience in the North Sea, the survey was fairly easy. I managed to find the J-tube flange and did a very thorough video inspection of it. There were a few bolts missing that the divers had either not bothered or forgotten to install, but in general, everything looked good. After my first dive, the young guys on board donned their lightweight helmets and installed any missing bolts. Because our job was complete, we left the area and headed for Venice, all of us looking forward to hitting the one bar in town.

After we had been steaming for a couple of hours, we received a disturbing radio message: the oil company had pressurized the J-tube and it was leak-ing. We couldn't believe it. We had just left the area, and everything was fine. They asked us to go back to check things out, so we turned around and in the moonlit night headed back to the platform.

What we discovered when we arrived was nothing short of astonishing: a bubbling cauldron of what I assumed to be natural gas, easily several hun-dred feet in diameter. The area was not gas and not seawater, but a gasified mixture. There was no way we could go in there. Ships float because they dis-place more volume and weight of water than they weigh. Ships cannot float on any type of gas. They will sink. If we took our ship into that area, it would sink like a stone. So we had to stay clear and figure out a plan.

Frenzied radio calls ensued. The oil company wanted us to find out where the leak came from. We told them that we would not under any circumstances go in there, but maybe, just maybe, we could stand off and send the Scorpi ROV in for a look to see where the leak was. Consequently, the ship anchored clear of the area, and we launched the Scorpi, hoping we had enough tether cable to reach the platform.

I slowly crawled along the bottom with the vehicle and managed to get close to the corner of the platform using the scanning sonar in the location where we had earlier inspected the J-tube. But the closer I got, the more ter-rifying the conditions became. It was not long before I had the ROV planted on the bottom just yards away from what appeared to be an underwater volcano. From what I observed, there was gas spewing out at high pressure from within the rig, and it wasn't from our J-tube. The entire area had been transformed into an underwater blizzard, where chunks of mud were being blown up from the bottom and raining down like hail. None of us had seen anything like it. The ship's crew didn't like it either because there was no way to know whether the leak could intensify and gasify the waters around our ship, sending us to the bottom in short order.

Somehow, I managed to find our J-tube and flange. It was intact and not leaking. But to my amazement, the pipe was no longer sitting on the bottom; it was now at mid-water, suspended at least 10 feet off the seafloor. The high-pressure gas had simply carved out a huge portion of the bottom, sending all the sediment far up into the water column, which was now raining down all over the place.

They asked me to get closer, which I did not really want to do. But I did. Before I could react with the Scorpi's thrusters, the vehicle was swept up into the vertical cauldron of gas and carried straight up toward the surface and into the rig. I estimate that we went from the bottom at almost 500 feet to the surface in less than thirty seconds. Nothing I tried helped at all. The vehicle was out of control and being tossed around and upward by forces that were impossible to counter.

The Scorpi popped up on the surface close to the platform. I could see the lights of the ship and tried to drive back on the surface but ran out of cable; the vehicle's tether was fouled somewhere down deep within the rig. I dove back down, following the cable, and tried for hours to free the vehicle, but it was impossible. Before we departed the area, we had to do what all ROV pilots hate: disconnect the topside end of the umbilical and tie it off on the side of the platform for another day. Weeks later, after I was back at the Eastport shop, I heard that divers had finally secured the gas leaking from the interior of the platform. All that was left of the ROV was a cable snaking down from the rig into the bottom. So much sediment had been blown up from the gas leak that it had buried the Scorpi. That was the end of that.

A couple of years passed after the Sea Plow job as I continued to work at Eastport. Craig Mullen, the president of the company, eventually sold the firm to Oceaneering International, a much larger undersea services company. Most of us stayed on working in the company's Advanced Technologies Division in Upper Marlboro, Maryland.

While we were doing our usual maintenance work, TWA Flight 800 went down in the Atlantic Ocean on July 17, 1996, at approximately 8:31 p.m. EDT. The aircraft crashed shortly after takeoff from John F. Kennedy International Airport. Terrorism was immediately suspected, especially because many eyewitnesses to the crash recalled seeing a streak of light traveling from the horizon to the fireball of the exploding jet. All 230 passengers died in the crash, which is the third-deadliest aviation accident in US history.

While the National Transportation Safety Board eventually concluded that a wiring short circuit in the plane's center wing tank caused the explosion, the missile theory has persisted. Much of this speculation has been pushed by author Jack Cashill, who has written books and articles on the possibility that an errant US Navy missile brought down the jetliner. However, the prevailing conclusion from the NTSB is generally accepted as the cause of the accident. It is not the purpose of this narrative to debate what happened to TWA Flight 800, but simply to describe my participation in the search and recovery efforts.

Our support of the salvage of TWA Flight 800 began on the USS *Grasp* (ARS-51), a combat salvage ship based out of Little Creek, Virginia. The ship and her crew had just completed a five-month deployment but were sent to the crash location anyway (standard protocol for such events). Our job was to find the flight data and voice cockpit recorders and assist with finding and recovering passengers from the flight.

The vehicle we used was the MR-1, a modified Sea Rover built by the late Chris Nicholsen of Deep Sea Systems and owned by the US Navy. It was a hand-portable vehicle that we launched from the port side of the ship (the left side of the ship looking forward) using a small davit. The vehicle was fitted with an acoustic navigation beacon so we could track it while underwater.

The vehicle's topside consoles consisted of two racks that were situated near the ship's machine shop close to a large welding table and drill press. We didn't have a dedicated control van for the ROV. The consoles contained numerous support equipment such as video monitors, a high-voltage DC power supply, sonar displays, and a DVR. The pilot drove the vehicle using a small handheld controller—just a small box fitted with a joystick and several buttons. It was a good-driving vehicle and was fitted with a very simple three-function manipulator that we could use to recover small objects.

The aircraft's flight data recorder was found during my first day on location, almost by accident. I cannot recall whether anyone had done an acoustic search for the black box, and the Navy's formal salvage report does not mention the circumstances of how it was found. My memory recalls that during a survey of the main debris pile underneath the USS *Grasp*, a Navy diver simply stepped on the recorder with his boot while the area was illuminated by the lights of our MR-1 ROV. After looking down, he saw the familiar orange color of the box and simply picked it up by hand. Flight data recorders and cockpit voice recorders are both fitted with acoustic beacons (37.5 kHz), which can be detected at a range of over a nautical mile. Unfortunately, these beacons are many times destroyed or rendered inoperable by the crash. In addition, the beacons need to be maintained, though in many cases they are not. Maintenance is simply removing the beacon every five years and replacing it with one that has a fresh battery. Given all the procedures followed by major airlines, I still find it inexplicable that these are not changed out on a regular basis. For example, the acoustic beacons fitted to Malaysia Airlines Flight 370, which disappeared in 2014, were all out of date. In some cases the airlines cannot even confirm the operating frequency. New regulations require that both a high frequency (37.5 kHz) and low frequency (8.8 kHz) acoustic beacon be fitted to commercial aircraft. The low frequency beacon allows a much longer range of detectability.

This turned out to be a grueling operation for various reasons. First, we were berthed in the crew recreation room. The entire crew hated us for that because we were living in their off-shift hangout. It was not an ideal situation for twenty-four-hour operations. Our crew were living and sleeping in this one room twenty-four hours a day. With the lights on, you could see piles of luggage, clothing, towels, and work boots scattered around several temporary steel bunk beds. There was also a crew training area with computers in the back of the room. Consequently, people were always coming in and waking us up so they could do training on a computer. In addition, every morning at about 9:00, a seaman would enter the room dragging a fuel hose to sound the diesel fuel tank and fill it as needed. If that wasn't enough, there was a loud-speaker in the room and, again, every morning, it would erupt with the orders

"Sweepers . . . sweepers . . . man your brooms." We eventually stuffed a T-shirt into the speaker, but it did little to help.

The work was intermittent and not pleasant because our main job was to find deceased passengers and allow the Navy divers to follow the vehicle's umbilical down to the location and recover their remains in body bags. Once a bag with the remains inside was lifted to the surface using a diver's stage, they would cut a hole in the bag using a knife to drain out all the water and bodily fluids, then place everything in an outer body bag. The orange bag I used for dirty laundry had the biohazard symbol on the outside. On a regular basis, it was normal to see a dozen or so crash victims stacked on the stern of the *Grasp* like cordwood. It was depressing.

Prior to some operations, like TWA 800, the ship would get a visit from a Navy chaplain, who would counsel us on the job before us from the standpoint of mental and spiritual health. These visits were well received, but there was no denying that there were things we were going to see that would forever be burned into our memories. A person cannot easily unsee images of humans who have been ravaged by the violence of a high-speed airplane crash. The chaplain's best advice to us was to not look at anything we didn't have to. That was easy for him to say, of course. There was no way for us to not see such things and do our jobs at the same time.

I distinctly remember thumbing through the wallet of a young girl who perished in the crash. Inside were her passport, credit cards, and so on. I imagine that she never would have dreamed this would be her fate.

The video monitors in the machinery room (where we controlled the ROV from) were there for anyone to see if they wanted to. I remember hovering the vehicle near the body of a young woman, naked from the wind blast, with her body still held in by a seat belt, when a female crew member happened to walk by. She left with her hand clasped over her mouth in shock. But we had to look at such images every day. I cannot believe it did not affect our mental health; you can get PTSD from situations other than combat. There is nothing good about seeing fellow humans being eaten by crabs on the bottom of Long Island Sound.

TWA Flight 800 was not like Air India Flight 182, where the crash victims were over a mile underwater. We did not even recover bodies from that crash. But on TWA we did, all that we could find. There was so much organic material in the water column that it was difficult to see underwater at times. The Navy divers told us they could smell the decomposing bodies through their helmets. The stench filled the ship. Once, I found part of a skull, which I brought to the surface to be retrieved by a diver. Another time, I noticed a human spine nestled within the interior of one of the airplane's tires. All we could do was point these things out to the corpsman and let them deal with it.

At one point in time, we were visited by a congressional delegation that elected to observe the operation. I was driving MR-1 at the time and had to be careful to not allow the camera to come within sight of a deceased passenger. A member of the delegation commented that she would not want to play a video game with me.

During an especially odd cookout, everyone on the ship had burgers and dogs on the bow. At the time, there was a stack of body bags on the fantail. Happiness and life on the bow; death on the stern. It was a strange juxtaposition and I felt guilty. But I ate my burger.

The captain of the *Grasp*, Cmdr. Orr, had his ship moored over a large pile of wreckage, and nothing was going to make him move. It was his pile, and he wanted all the glory and commendations that came with it. During one dive, we smelled smoke drifting throughout the machinery room. There is nothing more dangerous for a ship at sea than a fire, and we had one. The boom cranes on the *Grasp* were old, and it turned out that the slew motor for the aft crane had failed and caught fire. All of us had to mass at our muster points while the crew donned their turnout gear and breathing apparatuses and got the fire under control. The ship's electrician managed a temporary repair, so when the admiral came on board the next day, Cmdr. Orr could demonstrate that the crane was still working, and he could stay on top of his wreckage.

It is difficult to remember the size of the wreckage pile we were working on, but it was easily 75 to 100 feet in diameter and rose roughly 10 feet into the water column. Contained within this pile of debris were miles of wiring, passenger seats, assorted structural components, and, of course, human remains.

The young Navy divers, to my recollection, were a combination of Explosive Ordnance and Disposal technicians, SEALs, and those from Mobile Diving and Salvage Unit 2. They would investigate what we found by descending in the diver's stage using surface-supplied air, or diving from a small rigid-hulled inflatable boat on scuba. They had a hard job because the water was very cold and the Navy was pushing the air tables. By pushing the tables I mean that they were at the extents of what were considered safe bottom times. But even with that, people were going to get bent (decompression sickness).

They were a good bunch of lads with their heavily tattooed bodies and pierced penises (some of them anyway). They were doing air diving with surface oxygen decompression, meaning that once they came out of the water, they had only a few minutes to get into the deck decompression chamber, get pressed back down, and start breathing oxygen with an oral–nasal mask. The worst part was that the *Grasp* did not have sufficient berthing to keep them all on board. Consequently, many of them were berthed on a nearby Navy logistic support ship. There was nothing wrong with this unless one of them developed symptoms of the bends. In that case, they had to be hauled back to the *Grasp*, sometimes in the middle of the night, and undergo a treatment regimen of pure oxygen. I remember them coming out of the water, being helped by two beefy guys, one on each arm, and escorted to the deck chamber. They always had a deer in the headlights look about them. That's what cold underwater work does to people. I think during the early phases of the operation two or three divers experienced decompression sickness every week.

Every day it was pretty much the same deal: We would dive the ROV, explore the immediate area under the ship, and report when we found human remains. On one occasion, I was far out from the ship, maybe 75 meters or so, and found a dead passenger. There was not much left, just bones and tissue encapsulated by clothing and a yellow belt. I called him "the man with the yellow belt." I notified Cmdr. Orr about my find, and he said it was too rough to put divers down. I dutifully marked the spot, hoping to be able to return later. Unfortunately, we were forced to break out of the moor due to incoming bad weather—a nor'easter coming through. The *Grasp* pulled up anchors and headed back to Staten Island to wait out the storm.

The city of New York was very good to us, giving us rooms at the World Trade Center Marriott. It was a welcome break as we took the Staten Island Ferry to the city for some well-deserved rest. I spent my time enjoying the wine supplied with the room and checking out the local music talent at a bar near the Bowery. Every once in a while I would run across fellow shipmates getting soused like me. It was a brief but welcome respite from all the death we had witnessed.

After the storm passed through a couple of days later, we headed back out and once again established our moor over the wreckage pile. I purposely dove the ROV and went back out to find the man with the yellow belt. Unfortunately, when I reached what I was sure was the right area, he was gone—blown into atoms or buried by the storm. There was nothing left. That was the case for most of the small components of the aircraft, but the wreckage pile was still there, and we worked it.

Later, Jimmy Buffett came to the area to give a concert for the people working on the recovery operations. Our boss didn't allow us to attend, so we watched the entire Navy crew go ashore and have a good time while we steamed in anger on the ship.

Eventually, the Navy decided to lift an intact wing from the Boeing 747. This was obviously a big job, and it took a lot of time for the divers to do the rigging. But after some hard work they got all the lines hooked up, and before long, we had the entire right wing from TWA Flight 800 resting across the *Grasp*'s fantail. The wing was so long that the ends, one of them still attached to a section of fuselage, protruded out from both sides of the ship. But now they had a problem because it was too large to be transported to shore on the small auxiliary craft they used for smaller wreckage and bodies. So, the wing had to be cut into sections. They first tried to accomplish this with a Sawzall, an electrically operated reciprocating saw. That didn't work very well. Eventually, they were forced to use a Broco underwater exothermic cutting rod, which used a tube of magnesium, fed by pure oxygen on the inside, to burn through the wing. These tools can cut through concrete and made short work of the wing. It was not just the wing they recovered, but also a section of the fuselage with several burned windows attached. But they managed to get the wing offloaded and sent to shore, albeit in sections.

After some six weeks on the *Grasp*, I was reassigned to the USS *Grapple* (ARS-53), a sister ship to the *Grasp*, to take charge of the Deep Drone ROV system during the next phase of the recovery. Deep Drone was a much larger vehicle, powered by both hydraulic and high-voltage electric thrusters. It also had two hydraulic manipulators, enabling us to recover wreckage without the help of divers. Our Oceaneering International team was well seasoned and could function almost autonomously while I directed the vehicle's movements from the bridge using a large plotting chart and a video feed of our acoustic navigation system. Like the *Grasp*, we were held in place by mooring lines, but not above any single wreckage pile. Our job was to clear the areas farther out from the primary crash location.

Working on the *Grapple* offered a sharp contrast in the way the ship was managed. Unlike the *Grasp*, the *Grapple* was a happy ship with a good man in charge, Capt. Robertson. He and I had discussions on the bridge and established ground rules for what was practical and what was not, given the technology at our disposal. I convinced him that if we tried to pick up every single part of the wreckage using Deep Drone, we would be there for years. Consequently, we mutually agreed to concentrate on wreckage larger than a breadbasket, which we would place into a bottom-deployed recovery basket for eventual recovery. The real focus of our search was the electric fuel pump from the center fuel tank. If we could find the pump, it might contain valuable forensic evidence to support the theory of why the Boeing 747 exploded in the first place.

By October 31 we were still combing the bottom for the elusive fuel pump. To recognize the day, we dutifully strapped a plastic jack-o-lantern to the side of the Seahorse 1 launch and recovery system. I always found it creepy to be over the site of a major airplane crash on such days.

The final phase of the TWA Flight 800 search was to bring in the scallopers, who dragged their steel dredges across the seafloor, penetrating deeper into the bottom and scooping up anything large enough to be snagged in the process. Our role was simply navigation and directing the drag paths the ship was following. I was fortunate to miss out on the activity because it involved working on small fishing boats. The government personnel assigned to oversee this work were from both the NTSB and the FBI since this was a criminal investigation.

My favorite story from this operation was relayed to me by fellow team members who worked on these boats. Apparently, sometime during the process, they dredged up what, to the uninformed (i.e., the FBI), appeared to be a set of fins from a surface-to-air missile. The bureau's agent on board was very excited at the find and was ready to call headquarters to tell them of the amazing discovery. That is, until one of our people casually informed him that the fins were from a towed side-scan sonar, probably lost by one of the National Oceanic and Atmospheric Administration survey ships.

Once trawling operations were completed, the search for TWA Flight 800 was concluded.

After TWA Flight 800, there were numerous other operations I participated in, such as a McDonnell Douglas F/A-18 salvage off of Norfolk, a Boeing E-3B recovery in the Mediterranean near Haifa, Israel (where we also discovered an intact Bristol Beaufighter ditched in 1945 in 5,000 feet of water), a Sikorsky HH-60 helicopter salvage, and sea trials with a 35-metric-ton underwater sea tractor for the Navy out of Cheatham Annex in Virginia using the USNS *Zeus*, a massive cable ship. Then followed the first live video transmitted from the RMS *Titanic* using the Magellan 725 ROV and the finding of the *Liberty Bell 7* Mercury spacecraft using the same system.

10

1999: They Said It Would Never Be Found — *Liberty Bell 7*

J ust exactly who were "they"? Take your pick, from NASA, to the US Navy, the Smithsonian Institution, former astronauts, and most of my peers in the undersea industry. Nobody thought *Liberty Bell 7* could be found. It was too small, the waters were too deep, and no one seemed to really know exactly where it sank. I had one and only one hope: that if we could get a side-scan sonar within range of that capsule, the sound would bounce off it like a ping-pong ball, revealing its location.

I had been visualizing an expedition to find, and hopefully recover, Virgil I. "Gus" Grissom's *Liberty Bell 7* Mercury spacecraft for over a decade. For me personally, the reasons were twofold: First, I thought it would be fun. For once, after spending over a decade inspecting oil rigs and pipelines and burying telephone cables, I would be able to use the technology to do something that I wanted to do. In other words, I wanted to demonstrate our capabilities while doing something historic. Second, as a ten-year-old kid living in Florissant, Missouri, I was aware of Grissom's flight in 1961. I didn't remember much about it due to my young age, but I was especially impressed with Grissom as an astronaut. They were building Mercury spacecraft not far from my home at Lambert Field, the location of McDonnell Aircraft. If *Liberty Bell 7* was still intact, why not find and raise it?

My first thoughts of attempting such a brazen feat were during the Air India salvage in 1985. *If we can raise wreckage from over 6,000 feet underwater, why not* Liberty Bell 7? I pondered.

There were a lot of reasons why not; I just didn't understand them at the time. First, we needed a submersible vehicle, manned or unmanned, capable of reaching the depths where the spacecraft was lost, estimated at over 15,000 feet underwater. They didn't exist in 1985. However, I felt confident that, one day, they would be a reality. There were some manned vehicles, such as the *Trieste* bathyscaphe, that had even exceeded those depths, but the *Trieste* was not operational, and no other submersible at the time was a practical option.

The second problem was where to search for the capsule. *Liberty Bell 7* was lost on July 21, 1961, following a successful fifteen-minute suborbital flight. This was the United States' second manned spaceflight, following a similar mission by Alan Shepard in his *Freedom 7* capsule on May 5, 1961. Launched by a Redstone rocket, Grissom and his capsule reached an apogee of about 115 miles above the Earth, eventually landing in the Atlantic Ocean 320 miles from Cape Canaveral. It was after landing that the problems began.

At that time, NASA did not have a proper postlanding procedure telling Grissom what to do after splashdown. The Mercury spacecraft Grissom was piloting was the first to be fitted with an experimental explosive exit hatch. The hatch actuation mechanism was a simple device consisting of a single plunger or button that mechanically fired two percussion caps, setting off a ring of mild detonating fuse surrounding the hatch opening. Following his landing in the Atlantic Ocean and with the prime recovery helicopter hovering nearby, Grissom decided to get ready for pickup by removing the protective cap from the plunger as well as a small safety pin. In doing so he had effectively armed the hatch, which only needed the plunger to be depressed to set it off. During his postflight debriefing on Grand Bahama island, Grissom denied touching the hatch plunger after arming it.

The primary recovery helicopter was *Hunt Club 1*, piloted by Marine Lt. Jim Lewis, with John Reinhard, his copilot, kneeling in the open door with a long recovery pole. The shepherd's hook end allowed Reinhard to connect a lift wire to the top of the capsule. By his own account, Grissom was "minding his own business" when the hatch detonated, sending a wall of sea-water into the capsule. According to Reinhard, as soon as the end of his re-covery pole reached the top of *Liberty Bell 7*, he observed an arc of static elec-tricity running from the pole to the capsule. It is well-known that hovering helicopters build up a potential of static electricity that must be discharged to Earth using grounding cables. Remastered 16 mm footage of the encounter by Andy Saunders and George Leopold proves that Reinhard's recovery pole was touching the top of *Liberty Bell 7* when the hatch blew. Therefore, if the static electricity theory is correct, it would have made no difference whether or not the hatch was armed; it still would have detonated. All that static po-tential cared about was the Mercury fulminate inside the percussion caps. The astronaut was lucky to escape as Lewis's helicopter managed to snag the cap-sule while Grissom, foundering in the ocean with his spacesuit filling with water, was rescued by a second chopper.

For five long minutes, Lewis tried to save the spacecraft. Unfortunately, over time, his engine began to overheat and he observed indications of an imminent engine failure. Reluctantly, he radioed back to the recovery ship, the carrier USS *Randolph*, that he was jettisoning the spacecraft. *Liberty Bell 7* splashed into the ocean on her side, filled with water, and sank out of sight, not to be seen again for thirty-eight years. Following her loss, NASA aban-doned the helicopter recovery method, instead relying on shipborne cranes to lift their Mercury spacecraft from the ocean after they were fitted with a flotation collar.

In 1961, there were no satellite navigation systems, and LORAN reception in the splashdown area was poor. Maybe NASA had managed to track the spacecraft with radar or some method unknown to me. I would have to de-termine whether it was possible to pinpoint the capsule's likely location ac-curately enough to locate it using a towed side-scan sonar.

As I wrote in my 2002 book *Lost Spacecraft*, most of the technologies we take for granted today simply didn't exist in 1961, and this fact was going to affect whether the project was even feasible:

> In every way, the world from which Gus Grissom would soon be launched
> was far different than today. Computers were massive contraptions that filled
> whole rooms, like the IBM 7090 mainframes at the Goddard Space Flight

Center. Calculators? There were slide rules. There was no such thing as satellite navigation because there were no navigational satellites. The aircraft carrier USS *Randolph*, a veteran of the Second World War and the prime recovery ship, fixed their position in the splashdown area using Loran A (a form of radio navigation) and sextants. Cable TV was something you used when you were in the middle of nowhere and had to put a quarter in a hotel room TV set. The satellite clock inside of his Mercury capsule was hand-wound and the earth path indicator consisted of a small plastic globe rotated by dozens of small brass gears. Grissom's Redstone booster was almost a direct copy of the German Army's 1945 V-2 rocket and used the same fuel; it didn't even have a gimbaled exhaust nozzle, instead relying on graphite vanes for guidance. In fact, the escape tower used on the later Apollo Command Module had more thrust. The Beatles were still some unknown British pop band in Liverpool and if you wanted money, you had to go to a bank during business hours instead of an ATM machine. People bought 33⅓ speed vinyl High Fidelity records instead of CDs and had never heard of 8-track or cassette tapes. Television was black and white, and the NBC television footage taken of Grissom's splashdown were kinescopes. There was no calling Houston and saying ". . . we have a problem," because the Johnson Space Center didn't exist.

Another thought I had was the probable condition of the spacecraft after decades on the seafloor. Would the capsule even be intact? Was it now just a pile of corroded metal? Was it even worth recovering? I didn't know the answers to any of those questions but was determined to find out. All of this was boiled down to three critical questions:

1. Did the technology exist to find and raise *Liberty Bell 7*?
2. Could a search area be developed that likely contained the target?
3. Would the capsule's condition be good enough to justify recovery?

A final question, equally important, was who was going to pay for it all. This was bound to be an expensive undertaking, and I didn't know anyone with millions of dollars in their back pocket. I would have to find them and convince them that this was a worthy endeavor. This became my mission in life for the years stretching from 1985 to 1999.

By 1998, I had at least achieved one dream in life: diving on the RMS *Titanic*. I managed to convince Oceaneering International that I would be a good addition to the Magellan 725 team given my piloting abilities and knowledge of the wreck. Fortunately, they agreed, and I was able to create some stunning video of the bow section, thanks to the incredible camera system developed by Woods Hole Oceanographic Institution. It was during our mobilization that I received a phone call out of the blue from the Discovery Channel expressing an interest in *Liberty Bell 7*. I honestly didn't think it would go anywhere but jumped at the chance just the same. That was in August of 1998. Incredibly, by January of the next year, Discovery had agreed to fund the expedition and signed contracts were in place between them and Oceaneering, as well as the Kansas Cosmosphere and Space Center, who agreed to restore the spacecraft if it was found and raised from the abyss. I was designated the expedition leader to oversee the operation. I also convinced Oceaneering to give me a leave of absence to manage the project.

We all agreed on an April departure from Port Canaveral using the Ocean Explorer (OE) 6000 side-scan sonar to search for the spacecraft, and the same Magellan 725 remotely operated vehicle (ROV) used on the *Titanic* for recovery. Veteran Steve Wright was in overall charge of the equipment, with Ron Schmidt supervising the night shift. I wished their positions were reversed. Unfortunately, Ron angered Oceaneering management by leaving the commercial division and transferring to the Navy's undersea operations program to get a Magnum ROV ready for issue. But at least he was on the job because he and Mark Wilson, another fellow technician, knew the systems like no one else. After a few days of loading equipment, we left Florida in the afternoon on April 19. However, not everyone was pleased that the expedition was happening. In his book *The Unbroken Chain*, Guenter F. Wendt, NASA pad leader, writes:

> As Newport and his team prepared for the April castoff from Port Canaveral, Gus' widow, Betty Grissom, began raising hell with the press. She said she hoped that they would not find it and was "disgusted" that no one had consulted her about the recovery. NASA and the Smithsonian Institution had both carefully studied Newport's plan and given it their approval. Personally, I don't know what business it was of hers. . . . The Kansas Cosmosphere

would do the restoration and oversee its exhibition. . . . Apparently, she believed that the Cosmosphere, at some point, had tried to get Grissom's GT-3 spacecraft moved from her hometown of Mitchell, Indiana, to the Cosmosphere's elaborate museum in Kansas. . . . All the newspapers were printing the story. Now she did not want the Cosmosphere to have anything to do with *Liberty Bell 7*.

By April 21, we had completed testing with the OE 6000 and had started our first real dive in the area, a block of ocean floor in the Blake Basin 3 nautical miles wide and 8 miles long. It was a pitifully small chunk of ocean in which to find something lost thirty-eight years ago, but it was all we could afford. At a depth of 2,870 meters, however, all hell broke loose. We immediately lost the telemetry connection with the sonar, and after dragging the vehicle back on deck we discovered about a quart of seawater in the port electronics bottle. Mark and Ron dove into repairs straightaway with their usual vigor, while Steve smoked and complained.

Incredibly, they had the towfish working again later that day, but in their haste they had neglected to fully tighten a pressure fitting connected to the system's internal depth sensor. That single mistake nearly doomed our expedition as seawater leaked into the electronics bottle, frying the fragile electronics. The sonar was not repaired until April 24 after doing a pressure check down to 1,000 meters. They did this without the electronics chassis installed in case we still had leaks. Ron and Mark had worked tirelessly in their efforts to fix the sonar's electronics. Now it was time to see if their efforts would pay off.

By April 30, we had completed our search lines. It took hours to get the OE 6000 to depth, and anywhere from eight to twelve hours just to turn the ship at the end of each pass. It was a painfully slow process. Our search procedure was to do primary lines, so that the 1,000-meter swath of the sonar (500 meters on each side) overlapped. Then, we would inspect the nadir, or the area directly underneath the sonar, by doing secondary, or fill-in, lines to make sure the capsule was not hiding in the weeds under the OE 6000. Although we almost killed her twice, the OE 6000 turned out to be a good acoustic servant, churning out hours of sonar data, some of it hopefully containing a reflection from *Liberty Bell 7*.

As it later turned out, Grissom's sunken capsule was detected during the next to last fill-in line, only a half nautical mile inside of the western border of our search area. It was identified in our line records as target 71 and was a good reflector of sound on a high point in the terrain—the proverbial acoustic ping-pong ball I had been hoping for. Our problem was that the chirps of low frequency sound from the OE 6000 were bouncing off a lot of other things, such as the wreckage of a crashed twin-engine Piper aircraft known to be in the area. Who could have imagined that in the expanse of the North Atlantic Ocean, an airplane could crash right into the middle of our search area? Incredibly, it did.

Before diving the Magellan, myself, Richard Daley, and Mark Wilson spent hours reviewing the eighty-eight individual sonar targets we had detected. This analysis was normally done in the evenings in the 20-foot control van with the air conditioning going full blast as we drank gallons of coffee and chain-smoked (Richard and I, anyway). Mark was a clean-cut-looking man, highly intelligent, and adept at hacking into satellite television networks. Richard was a husky Texas boy who spoke with a slow drawl and had decades of underwater survey experience. For each target, our technical banter would go something like this:

"OK, so we have this contact, ping number 3138. Not a big reflector. Small target on the side of a hill," Mark said after glancing down at one of the target sheets. The image on the video screen was that of a tiny splotch of white-and-yellow pixels with a couple of red ones in the center.

Richard quipped up, "Yeah, but see that trail of soft contacts coming down from the northeast? A definite geological feature. It looks as though they are tied to the target because they lead right up to it. I think what we have here is geology. Maybe some compressed sediment or mud."

After listening to their comments, I added, "The other thing is that there is no shadow. What we're looking for should definitely have some shadow. I agree with Richard—probable geology." We gave this target only two stars, our subjective method of rating sonar targets.

Glancing down at the target sheet, Mark pulled up another contact on the screen of the navigation system's video monitor. "Next one is target 71—hard reflector on a rise. I like this one." This contact was a definite hard ringer, or an excellent sound reflector. There was no geology around it and it was an isolated target. These were good things. It simply lit up the sonar due to its high reflectivity.

"Man, I really like this one," I noted. "Look at the way the sound bounces off it, plus there is no geology connected to it. In addition, it is in the right spot for me, near the last corrected radar tracking position." Target 71 was a massive smear of white-and-yellow pixels on the screen with a deep red center.

Placing his face right in front of the monitor, Richard studied the image intently. "Yeah, this is a good one, I would give it four stars."

Mark glanced down at the sheet. "Four stars it is," he confirmed as he scribbled in the notebook.

When we were done with our evaluations, we managed to whittle down the eighty-eight targets to only fifteen worthy of further investigation. Still, those were a lot of contacts to individually inspect.

On May 1, 1999, we managed to get the Magellan 725 ROV to the bottom and do what "they" said was impossible.

At 4:17 in the afternoon, I sat with my face in my hands in the Magellan ROV control van. Steve was driving the ROV as usual. "Bottom in sight at 4,720 meters," he announced.

I looked up momentarily and saw that the Magellan was stirring up sediment in front of a small rise on the bottom. Small particles drifted past the vehicle's camera as the HMI lights gave the image a ghostly blue hue.

Mark, who was operating the scanning sonar, glanced at me and said, "Sonar's down." *Shit.*

Steve planted the ROV hard on the bottom with the vertical thrusters. "OK, we're going to have to recover. Plus, it's getting too rough."

I got up from my perch on an upturned 5-gallon bucket and glanced outside at the conditions. The seas were picking up and I could feel the ship's motion more now. I could smell the salt air and feel some slight spray on my face. I had not slept in eighteen hours and it woke me up a little. Thirty minutes later, Mark had roused Ron Schmidt from his rack and he was on deck working on the sonar. I stood outside on deck holding a flashlight for Ron so he could tweak some potentiometers inside the rotating junction box on the side of the cable reel.

"Any change?" Ron asked Mark.

"No, still the same."

Ron pulled off an optical fiber connector as I handed him an alcohol-soaked wipe. He cleaned the connector and squirted some compressed air from a small can into it, then took a deep drag from his cigarette and reassembled the connector.

"OK, what now?" he asked Mark.

"Hey man, it looks like it is trying. Let me change some settings . . . OK, I think it's good enough . . . We got it," Mark replied.

Again, I illuminated the area with my flashlight, while Ron attached the cover to the junction box with a socket wrench. When we finished, I turned a small fan back on to keep the surface electronics cool in the hot Bahamian sun.

Because the sonar seemed to be working and it was better to stay down, we started searching for target 71. The weather was worsening, and we didn't dare recover right then anyway.

Steve seemed unconcerned about how things were going as he clipped the top of sand waves using the Magellan ROV with much glee. We were stretching out the Magellan's soft umbilical on various compass headings in an attempt to locate this first target. As I watched from my perch on the paint bucket, anger built up inside of me. *What the hell does Steve care? It's not his expedition. He is not the person who's spent fourteen years trying to do this. It's not his money, but someone else's. He has nothing to lose. But I have everything to lose.* Those were the thoughts racing though my head at 6:03 p.m. on that day, on board the MV *Needham Tide*, 90 nautical miles northeast of Grand Bahama island, in rough seas, on the first of May in 1999.

I sat in silence and listened to the sweep of the scanning sonar. The Magellan ROV carved tracks through the bottom muck as Steve drove out on various headings, finding only a 2-foot-long section of rusty pipe. "There's nothing here," Steve declared, "Let's head over to the next one."

Before I could say a word, Mark, studying the screen of the scanning sonar, disagreed. "Well, why don't we make sure. Let's slide the ship over 50 meters to the west and properly clear the area."

Steve grunted and grabbed the microphone to call the bridge. "Bridge, op hut."

The wheelhouse responded, "Bridge."

"Can you hold this heading and reposition to the west about 50 meters?"

"Roger that."

AT THE PRESIDIO. A young Newport family in the early 1950s, with (*left to right*) mother Flora with the author on her lap, older brother Chris, and then Capt. Newport. The Presidio was a great place to grow up—until I fell off the top of a parking garage and landed on my head. (PHOTO FROM AUTHOR'S COLLECTION)

ARMY AVIATOR. An early image of Dad in the cockpit of a training aircraft. He always admitted that flying didn't come easy to him. Even so, he became a very skilled pilot with the US Army, specializing in helicopters. (PHOTO FROM AUTHOR'S COLLECTION)

KOREA. Dad excelled at flying his Bell H-13 helicopter ambulance in Korea and was the only pilot who flew at night for the 4077th MASH. By the end of his tour, he had completed over fifteen hundred evacuations from the front lines, earning the Distinguished Flying Cross along the way. (PHOTO FROM AUTHOR'S COLLECTION)

THE ARAKELIAN BROTHERS. Uncles George, Bobby, and Charlie (*left to right*) ran a produce shop on Fourth Street in Oakland, California, for over forty years. They all taught me what it means to have a work ethic, and I have many fond memories of helping them make their spinach, coleslaw, and carrot sticks. (PHOTO FROM AUTHOR'S COLLECTION)

DR. NORSUDA NEWPORT. Dad always told us that Norsuda was his brother, when in reality he was his uncle. It didn't matter as the two of them developed a very close bond while Dad was growing up. We always loved having Norsuda visit us because he taught us things about the local fauna. (PHOTO FROM AUTHOR'S COLLECTION)

DOWNRANGE. My favorite photo of my father because it shows the stress of combat as well as the pistol he used to save his life. His L-20 Beaver aircraft is in the background. This image is from his first combat tour with the Army Concept Team in Vietnam (ACTIV). Dad's last Army posting was managing the development of the Apache attack helicopter at the Hughes Aircraft plant in Culver City, California. He was the perfect choice. (PHOTO FROM AUTHOR'S COLLECTION)

PARATROOPER. My brother, Chris, during paratrooper training with the 82nd Airborne. Chris became disillusioned with the US Army while with the Berlin Brigade and left the service after five years. (PHOTO FROM AUTHOR'S COLLECTION)

IN ENGLAND. The author in the cockpit of a Lotus 51C Formula Ford racing car in 1970. The Jim Russell Racing Driver's School at Snetterton Circuit taught me a lot about driving on the edge. However, I didn't like the fact that none of their open-wheel cars had seat belts. (PHOTO FROM AUTHOR'S COLLECTION)

DRIVING THE ZINK. The author racing a Zink Formula Vee in the mid-1970s at Summit Point in West Virginia. The Zink was a great car, and I managed to set a lap record with it at Virginia International Raceway. (PHOTO FROM AUTHOR'S COLLECTION)

DIVING SCHOOL. The author wearing a Savoie air diving helmet at the Commercial Diving Center in Wilmington (Los Angeles), California, in 1977. CDC was a great experience and I finished second in class. (PHOTO FROM AUTHOR'S COLLECTION)

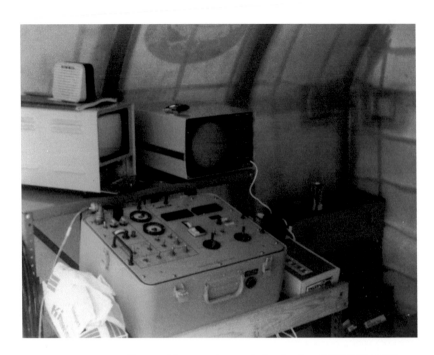

SCORPIO CONSOLE. The topside control console for the first SCORPIO ROV was incredibly primitive by today's standards. All it had was a small black-and-white video monitor, the topside sonar display, and a simple control box with two tiny joysticks. But it was very mobile and could be set up anywhere. (PHOTO FROM AUTHOR'S COLLECTION)

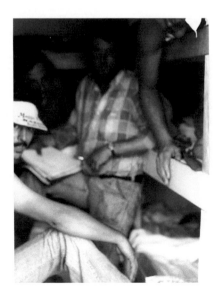

CRAMPED QUARTERS. *Left to right:* The author, Russ Austin, Glenn Tillman, and Chuck Collins. Our berthing area during the first job consisted of an 8-by-8-foot cabin, which doubled as the ROV control center. During a typhoon, the floor would be flooded in no time. (PHOTO FROM AUTHOR'S COLLECTION)

OUR GALLEY. The *Deepsea I* didn't have a galley, so we had to eat our meals on deck in the weather. Chuck Collins is seen exiting the single head on the ship, which doubled as our cold-water shower and was gravity fed. (PHOTO FROM AUTHOR'S COLLECTION)

ROV ON DECK. There was virtually no room on the fantail of the *Deepsea I* to work on the SCORPIO ROV. It was even worse when we were at sea. (PHOTO FROM AUTHOR'S COLLECTION)

NORTH SEA. One of several flare stacks attached to the main complex of the Ekofisk field in the Norwegian sector of the North Sea. The entire structure was over a mile in length, and we inspected all of it using a small TREC ROV. (PHOTO FROM AUTHOR'S COLLECTION)

NEKTON ALPHA. The author with General Oceanographic's Nekton Alpha manned submersible in 1978 during a pipeline inspection in the Gulf of Mexico. The small sub could dive to 1,000 feet and was part of a small fleet of such vehicles designed by Doug Privitt of California. (PHOTO FROM AUTHOR'S COLLECTION)

PREDIVE ON THE MV *IBIS 7*. One of our British shipmates conducting a predive on a Canadian TREC ROV in 1979. The conditions in the North Sea were reasonable in the summer—less so during the winter, when most worked stopped. (PHOTO FROM AUTHOR'S COLLECTION)

ROV LAUNCH. The SCARAB II just before launching into the waters near Ireland during the Air India Flight 182 salvage in 1985. As shown, it takes a lot of bodies to get a large vehicle into the water safely. (PHOTO FROM AUTHOR'S COLLECTION)

OVER THE SIDE. SCARAB II just before descending into the frigid North Atlantic waters in 1985. The large claw on the vehicle's right arm is what we used to puncture and attach lift lines to Air India aircraft wreckage. (PHOTO FROM AUTHOR'S COLLECTION)

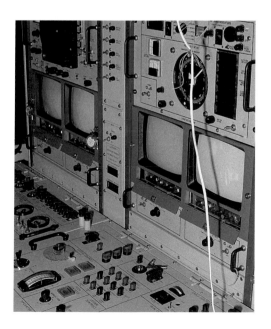

SCARAB II CONSOLE. The ROV's main operating console in 1985. The myriad of switches and dials made the vehicle work. It was not an easy system to pilot but very reliable for most jobs, sometimes staying on the bottom for as long as five days. (PHOTO FROM AUTHOR'S COLLECTION)

SCARAB PILOT. The author at the SCARAB pilot console wearing typical garb and a Bell Telephone StarSet communications headset. It was not always easy to stay awake during the long twelve-hour shifts. (PHOTO FROM AUTHOR'S COLLECTION)

VEHICLE CHECKS. The late Mark Proulx checking out the SCARAB vehicle before launch. Mark was a former US Marine who saw duty on the USS *New Jersey* off the coast of Beirut. We spent much time together in Ireland. (PHOTO FROM AUTHOR'S COLLECTION)

THRUSTING. SCARAB's four axial thrusters churn the water while maneuvering to stay in position during a launch. (PHOTO FROM AUTHOR'S COLLECTION)

ON DECK. The author on the deck of the CCGS *John Cabot* wearing a Mustang exposure suit and David Clark head-set. As the months of the Air India salvage dragged on into the winter, it became more and more difficult to stay warm. (PHOTO FROM AUTHOR'S COLLECTION)

GEMINI ROV. This vehicle was used to attach lift lines to all of the heavy solid rocket booster wreckage lost during the *Challenger* explosion. Gemini's powerful thrusters allowed us to work in the strong currents of the Gulf Stream. (PHOTO FROM AUTHOR'S COLLECTION)

WRECKAGE ON DECK. Morton Uloff of Stena Line directs the recovery of a large booster fragment. Most of the *Challenger* wreckage was heavily laden with several inches of volatile solid rocket fuel. (PHOTO FROM AUTHOR'S COLLECTION)

AFT SKIRT OF THE RIGHT-HAND BOOSTER. The author in front of the heaviest fragment of *Challenger* recovered, a 14,000-pound section of the aft skirt. It took three tries to get it on deck. Overall, we recovered about 30 tons of wreckage from the doomed space shuttle. (PHOTO FROM AUTHOR'S COLLECTION)

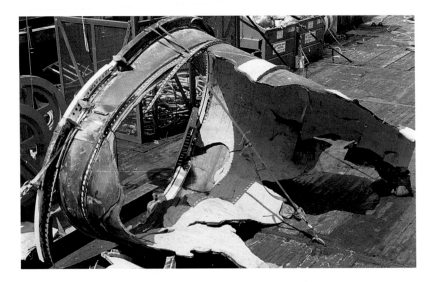

AFT SKIRT. Another view of *Challenger*'s right-hand booster. The edges of the wreckage were so sharp that they sliced through many of our wire ropes. I was finally able to
drag a heavy chain through the center of the wreckage using the Gemini ROV by flying
through the inside of the wreckage. (PHOTO FROM AUTHOR'S COLLECTION)

MANGLED TWA FLIGHT 800 DEBRIS. This chunk of wreckage was typical of what we
recovered from the aircraft. We used our small ROV, MR-2, to spot wreckage and
provide light to US Navy divers as they rigged the debris for recovery. (PHOTO FROM
AUTHOR'S COLLECTION)

POIGNANT REMINDER. The TWA logo clearly visible on a small section of TWA Flight 800. (PHOTO FROM AUTHOR'S COLLECTION)

LANDING GEAR. A partially destroyed main landing gear assembly from TWA Flight 800 with US Navy deck crew. During one of our recoveries, I discovered a human spine inside a tire from the plane. (PHOTO FROM AUTHOR'S COLLECTION)

TWA WING SECTION. Getting the left wing of the Boeing 747 on deck was not easy because it was too large to fit on the deck of the USS *Grasp*. It was eventually cut into sections using a thermal lance and transported to shore. (PHOTO FROM AUTHOR'S COLLECTION)

HUNT CLUB 1. The pilot of this Sikorsky helicopter, Jim Lewis, drags Gus Grissom's *Liberty Bell 7* across the seas in an attempt to get it back to the carrier USS *Randolph*. The spacecraft was jettisoned shortly after this image was taken. (NASA PHOTO)

OCEAN EXPLORER 6000. The OE 6000 was the deep-tow side-scan sonar we used to detect *Liberty Bell 7* in over 4,800 meters of water. It was not a new system, but the OE 6000 had found many deep-water shipwrecks in the past. (PHOTO FROM AUTHOR'S COLLECTION)

MAGELLAN 725 ROV. The Magellan was an antiquated system built from spare parts. Even so, it generated incredible images of the sunken *Liberty Bell 7* using cameras on loan from Woods Hole Oceanographic Institution. The vehicle was capable of reaching 7,000 meters and had a small 25-horsepower hydraulic power unit. (PHOTO FROM AUTHOR'S COLLECTION)

MOMENT OF DISCOVERY. The *Liberty Bell 7* spacecraft sitting on a pile of its corroded beryllium heatshield deep in the abyss. It had been there for over thirty-eight years and was in marvelous condition, with the lettering quite visible on the exterior. (DISCOVERY COMMUNICATIONS PHOTO)

RECOVERY. The author in the foreground as *Liberty Bell 7* is finally hauled on deck at 2:20 a.m. on July 20, 1999. It was a fourteen-year personal effort to find and raise this spacecraft. Success felt good. (DISCOVERY COMMUNICATIONS PHOTO)

USS *INDIANAPOLIS*. The heavy cruiser, identified as CA-35, in the waters off Mare Island in 1945 following a major refit. During her sprint to Honolulu carrying atomic bomb parts, the *Indy* set a record that still stands. (US NAVY PHOTO)

LAUNCHING THE SM-30 SONAR. Williamson and Associates deck crew launching our side-scan sonar during the first ever expedition to search for the USS *Indianapolis* in the Philippine Sea in 2000. While we didn't find the ship on that mission, we did learn quite a bit about the terrain near the Kyushu-Palau Ridge. (PHOTO FROM AUTHOR'S COLLECTION)

BOTTOM STRIKE. What remains of our TPL-40 pinger search vehicle after clipping the bottom during the search for Air France Flight 447. Unfortunately, we had little data on the water depth at the time, which led to the impact. We managed to beat the towfish back into shape and continue the search. (PHOTO FROM AUTHOR'S COLLECTION)

DIVING IN MIR I. The author entering the Russian Mir I submersible before diving on an unidentified nineteenth-century shipwreck, codenamed "Atlantic Target." I made two dives to 4,800 meters and it was an amazing experience. (PHOTO FROM AUTHOR'S COLLECTION)

ANATOLY. The head of the Mir program, Anatoly Sagalevich, driving Mir I during our first dive. He was a highly intelligent engineer and scientist who did much for the Russian program. (PHOTO FROM AUTHOR'S COLLECTION)

AT DEPTH. The author holding up the flag of the Explorers Club at depth in Mir I. I ended up with a massive bruise all along my rib cage from straining to see out of the small viewport to my left. (PHOTO FROM AUTHOR'S COLLECTION)

ATLANTIC TARGET. Mir I hovers off the bow of the wooden shipwreck at a depth of 4,800 meters. In the end, we recovered about 1,500 Spanish silver and gold coins as well as numerous artifacts. At the time, it was the deepest shipwreck ever explored. (WOODS HOLE OCEANOGRAPHIC INSTITUTION PHOTO)

LT. MIROSLAV "STEVE" ZILBERMAN. Steve's nickname was "Abrek," and he showed himself to be a very skilled pilot of his E-2C Hawkeye surveillance aircraft he flew off the carrier USS *Eisenhower*. He would eventually make the ultimate sacrifice while saving the lives of his crew. (US NAVY PHOTO)

WRECKAGE. The remains of the tail section of Zilberman's E-2C aircraft after recovery from over 3,000 meters in the Arabian Sea. The color scheme of the Bluetails is clearly visible on the vertical stabilizer. (US NAVY PHOTO)

SS *EL FARO*. The container ship shown in a file photo. The sinking was so violent that the top two decks were cleanly sheared off, along with the all-important voyage data recorder, which was eventually located and recovered by the US Navy. (MARINE TRAFFIC PHOTO)

CURV 21 ROV. This remote vehicle was used to survey the sunken *El Faro* and recover the voyage data recorder. A smaller ROV called Xbot was unable to enter the bridge due to the collapsed structure and cables. (US NAVY PHOTO)

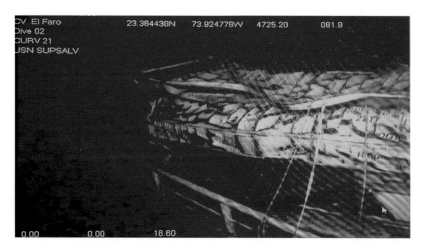

FLUID CONTAINER. This was the only container left on deck when the *El Faro* was surveyed. Most of the structure had collapsed due to the tremendous pressure at a depth of 4,600 meters. (US NAVY PHOTO)

SS *EL FARO* STERN. The name of the ship is clearly visible in this video still frame taken by CURV 21. Scores of buoyant mooring lines floated vertically above much of the ship, creating an excellent ROV trap. (US NAVY PHOTO)

RV *PETREL*. The bow of the research ship while docked in Guam before the start of the USS *Indianapolis* expedition in 2017. While the *Petrel* was not a new ship, Paul Allen spared no expense in outfitting it with the most sophisticated underwater and communications technology available. (PHOTO FROM AUTHOR'S COLLECTION)

USS *INDIANAPOLIS* IN TRANSIT. Quite possibly the last image ever taken of the *Indianapolis*, as it transited toward Guam in July of 1945. The photo was taken by a crew member of USS *LST-779* and helped me calculate the distance between the two ships as the *Indy* passed to the south. (FLOYD R. LAMBERTSON PHOTO)

REMUS 6000 AUV. The autonomous vehicle that eventually discovered the USS *Indianapolis* being launched from the port side of the RV *Petrel* in 2017. The Vulcan team developed an ingenious method to get the vehicle to depth very quickly: a massive descent weight attached to the nose of the AUV. (PHOTO FROM AUTHOR'S COLLECTION)

MISSION PLANNING ROOM. The author after a long day on the RV *Petrel*. On the left is a navigational planning monitor and Robert Kraft, director of marine operations for Vulcan. Paul Mayer, Vulcan's historian, is to the right. (PHOTO FROM AUTHOR'S COLLECTION)

THROWING DARTS. Sonar expert Gary Kozak (*foreground*) conferring with Robert Kraft during mission planning. The author is in the background. (PAUL ALLEN PHOTO)

Image Landsat / Copernicus

JAPANESE TROOP SHIP. This is a sonar image of what is probably the SS *Nanman Maru*, a Japanese ship that was torpedoed and sunk by the USS *Flying Fish* (SS-229) in 1943. This was the only other ship target found during the USS *Indianapolis* search. (PHOTO FROM AUTHOR'S COLLECTION)

USS *INDIANAPOLIS*. A side-scan sonar image of the *Indianapolis* as discovered by the Vulcan team. Fortunately, the target was in an area devoid of geology and was positively identified by an ROV. (PAUL ALLEN PHOTO)

AFT 8-INCH GUN TURRET OF THE USS *INDIANAPOLIS*. Typically, when a ship capsizes, these turrets fall off because they are held in by gravity alone. (PAUL ALLEN PHOTO)

EIJA LIPPO. Eija, my daughter Milana's mother, was a correspondent for the BBC World Service, Finnish Section, when I met her in 1992. Shown here on the ABC news set, she worked for the local Porvoo newspaper for many years. We stay in close contact. (EIJA LIPPO PHOTO)

MILANA NEWPORT. My daughter, Milana, is now studying chemistry at the University of Helsinki. Shown here inside the CURV 21 control center at Phoenix International, she lives in Espoo with her fiancé, Michael, and we communicate almost daily. (PHOTO FROM AUTHOR'S COLLECTION)

As I watched in silence, I could feel the vibrations from the engine's change in intensity and frequency while the helmsman adjusted the throttles to slip us over ever so slowly, struggling to keep the ship's bow pointed into the wind. In these conditions, one mistake could cause the ship to fall off badly and drag the Magellan almost 3 miles below with it. I heard a ballpoint pen roll off the console and my coffee cup rattle against the steel floor as the engines throbbed. Far below our ship, the Magellan's depressor swung like a pendulum above the pitch-black abyssal basin, dragging the floating ROV with it until all was settled. Finally, we were there as Mark studied the sonar screen intently. Steve set the Magellan back on the bottom in a cloud of sediment.

"I've got some small contacts up this rise in front of us," Mark commented.

"Maybe aircraft debris?" I suggested.

"Steve, see if you can move forward on this heading. These things look like they're about 25 meters ahead. Just a bunch of small targets," Mark said.

With that, the Magellan bounced up and down on a very tight tether, slowly climbing the hill.

After a few minutes, small chunks of material came into view. None of us had seen anything like them. They appeared almost as though they were chunks of ice, having a crystalline surface and random shapes and sizes. I stood up and stared at the video monitor, making out small marks in the sediment where they had obviously rolled down the hill like small snowballs. *Weird.*

Mark switched to a longer range on the sonar, announcing, "I've got a larger target behind all this stuff. About 20 meters away." Steve struggled with the hand controller as he tried to move up the hill. The vehicle was bouncing heavily and slowly clawing its way forward. It was a very tedious process.

I saw something begin to appear ever so slowly in the gloom but could not tell what it was. It looked to be a tall object, but at that moment it was unidentifiable. The Magellan continued to bounce up and down as it traversed the rise. More chunks of the mysterious material littered the area like trash.

I casually assumed that we were looking at a large chunk of aircraft wreckage. With the image bouncing up and down in front of my eyes, I slowly discerned the words "UNITED STATES," lit up like a neon sign in front of us. My pulse quickened and I felt my face flush. I muttered in total disbelief, "Oh my God. This is it!" As Steve forced the Magellan even closer, no one could deny what the four of us were seeing. We were the first people to see *Liberty Bell 7* in thirty-eight years, and for a discarded Mercury spacecraft, it looked pretty good. (My operational log records the time as 7:20 p.m. on May 1, 1999, with the notation "On Liberty Bell 7!")

The following description in Guenter Wendt's *The Unbroken Chain* captures what we saw:

> The spacecraft was in remarkably good shape. The black shingles on the sides looked almost brand new and the words "United States" were clearly visible in bold white letters. . . . At the top of the capsule, a large amount of corrosion blossomed out. At the bottom, the decayed remains of the heatshield seemed to form a bed of amorphous goo. As Magellan was steered around to the spacecraft's other side, the words "Liberty Bell 7" came into view. Even the white "crack" that Grissom had had painted on the side was still there. It was an exciting and marvelous sight.

Pandemonium erupted in the Magellan van as everyone tried to cram their bodies inside with director Peter Schnall, a soundman, and our website reporter. All hands wanted to see our find. I was still stunned at our luck, knowing full well that we were very, very lucky to get it on the first target. Unfortunately, with the crew focused on seeing the capsule, the conditions outside and ship movements went unnoticed. At some point in time, the ship's bow fell off, and with no one monitoring the overboarding sheave, a layer of wire began to strip off the outer layer of the Magellan's umbilical like the peel of a banana. We were quickly in deep trouble.

Ron Schmidt noticed the problem first and yelled inside the van to get someone on the crane because the umbilical was trashed. Bodies moved fast, coming on deck to discover a rat's nest of wire around the bull wheels of the traction winch, jamming it fast. Along with a few other guys, Ron leaped on top of the traction winch with a circular cutter and began to remove the damaged layers of plow steel. He was unconcerned with the sparks flying around him in all directions and dove into it.

Ultimately, all this work was to no avail because four minutes before midnight, the ship lifted up on a swell and the Magellan's umbilical snapped like a piece of thread. We were done. I lit another cigarette and told Steve to head to port. To stay out there now would be just wasting money. Ron and his crew secured the deck, the ship turned to the east, and we headed in. *What now?* I didn't know what now. Our only asset for recovering *Liberty Bell 7* was sitting with its target on the bottom of the Blake Basin in almost 3 miles of water. It was time to head back to land and regroup.

As word of *Liberty Bell 7*'s discovery shot around the planet, I was consumed by faxes, emails, and television interviews. I was nervous about these interviews at first and didn't especially like the fact that Discovery had announced publicly what we were attempting to do. Nothing like putting a bit of pressure on someone—namely me. But it was their money and I could not complain because at least we had an expedition.

I was not involved in whatever discussions Discovery had with Oceaneering regarding the way forward. All I knew was that we had indeed found the spacecraft, had a missing ROV on the bottom, and needed to recover both. The only reason to go back for the Magellan at all was because it had those wonderful Woods Hole cameras and lighting system on it. Other than that, it was not even worth recovering due to the expense. By then, the Magellan 725 was an old vehicle, by ROV standards, relying on a telemetry system developed in the early 1980s. But it had those cameras on it. What many don't realize is that the "7" in the vehicle's identification stands for 7,000 meters and the "25" stands for 25 horsepower. It was a 7-kilometer-rated ROV with only 25 horsepower; it was a good vehicle but getting long in the tooth.

At that time, 6,000-meter-rated ROVs were few and far between. Consequently, Oceaneering elected to construct a new vehicle, which would be named the Ocean Discovery. Their mistake was in not consulting with the one man they should have been talking to: Ron Schmidt. A major flaw in their design was using aluminum Impulse underwater electrical connectors. Ron told them that they leaked at depth, but he was ignored. The stainless steel versions were reliable, but of course heavier and more expensive. That caused us no end of problems when they started using the new ROV in July on the MV *Ocean Project*, a dynamically positioned support ship that Oceaneering insisted on using.

By July 6, I was back on location after joining the ship in Tampa, Florida. From the sixth to the eleventh, we aborted five dives due to various issues—many electrical, many hydraulic. It was not going well as we initially tried to locate the Magellan ROV so it could be recovered. What none of us knew was that our navigation contractor, John Chance, was using the wrong ellipsoid, putting us off position by hundreds of meters. When we initially searched for the capsule, we were navigating with the Clark 1866 ellipsoid because that was what NASA used in 1961. The ellipsoid defined the shape of the Earth in a particular way. Unfortunately, John Chance didn't send the same navigators out for our second expedition. They were using what's called WGS84 (WGS stands for World Geodetic System), which is commonly used today for most GPS locations. This put us off by 500 meters underwater, which is too far, given the range of our scanning sonar and cameras. As a result, we lost three days simply trying to relocate the capsule, time that I wanted to use to find and recover the explosive hatch. I had earlier reviewed all of the sonar contacts surrounding the capsule and had come up with three targets. I felt strongly that one of them was the hatch. But now I didn't have the time.

We didn't relocate *Liberty Bell 7* until July 11 and immediately began a detailed video survey so I could plan where to connect our recovery tools. I had designed these attachment tools back in the early 1990s. They were a simple design, looking much like an inverted vise. The difference was that the vise jaws were specifically designed to fit onto the Mercury capsule's escape tower retaining ring, which was rated for 20 gravities of acceleration. It was the *only* place to lift from. Earlier discussions about using a cargo net were deemed impractical.

During the hiatus between the search and recovery expeditions, the Smithsonian's Paul E. Garber Facility uncrated a Mercury spacecraft for me so I could determine exactly where to attach the tools. Many areas of the circumference of the escape tower ring had either small fasteners or small sensors, which would preclude a proper fit with the tools. I marked off areas of the top of the ring with red electrical tape and photographed everything.

With the Ocean Discovery ROV far below us hovering around the top of the capsule and my Garber photographs in hand, I directed Steve where to install the first tool. He was nervous, chain-smoking, and not dealing with the pressure well. Plus, he didn't like being told what to do, so I treaded lightly. As carefully as possible and trying not to collide with *Liberty Bell 7*, Steve attempted to align the first tool. It smacked the top of the capsule with an inaudible *thunk*.

The tool installation was a three-person job, with Steve flying the tool to the capsule, Richard Parks fine-tuning the tool orientation with the manipulator controls, and me studying the video monitor to make sure the clamp was in the right spot.

Pointing at the screen, I told Richard, "OK . . . See how it is not level with the top of the ring? Just give it a few degrees of clockwise rotation."

We watched the tool spin slightly.

"How's that look?" asked Richard.

"That looks decent. Retract the wrist a little. And Steve, give the vehicle a bit of reverse thrust to hold that rear jaw in place."

The reverse thrust made the ROV pivot slowly as the far jaw dug into the ring.

"Richard, start tightening up the clamp. And Steve, just keep a small bit of reverse thrust on to keep the jaw in place."

We watched the threaded shaft rotate slowly. Every so often, it would be side-loaded and bound up.

"A little bit of lateral thrust might help," Steve advised.

"Go for it," I responded.

Over a period of almost an hour, the slow process continued. Shaft rotated, then stopped, rotated, stopped. Eventually, the three of us could see it tighten up as Steve released the first clamp. It immediately self-aligned on the ring.

"I think if you just give it a final tighten with the jaws loose on the T-handle, we're good on this one," I told Richard.

"Roger that."

By 6:55 a.m. on July 13, after working all night, we had three recovery tools attached to the capsule and had installed the nylon lifting sling. With all of us exhausted from the concentration, before leaving the bottom we picked up the spacecraft's dye marker canister, then headed up to the surface, a five-hour trip.

After preparing *Liberty Bell 7* for recovery, we were chomping at the bit to actually *do* the recovery. The Ocean Discovery ROV, however, had other ideas, as one technical problem after another erupted out of nowhere: damaged vehicle compass, problems with the underwater cameras, lost telemetry links, electrical shorts in the ROV's hydraulic motor, damaged cables, leaking electrical connectors, sonar failure . . . Everything just kept going wrong, until finally we were forced to leave the area and transit to Marsh Harbor on Great

Abaco island to pick up a part for the topside sonar system. Simply getting off the ship for a few hours was a welcome respite from the stress of the operation.

Just after midnight on July 19, we were floating over the capsule once again, almost 3 miles below our ship. Once again, the ROV made sure we knew who was in charge when a water alarm forced us to recover again.

The steps required to lift the capsule to the surface were not complicated but needed to be done in sequence. Now that we had three recovery tools and the sling on the spacecraft, we had a single point from which to lift. If our ROV kept working, on the next dive we would sling a large aluminum spooler underneath the vehicle and transport it to the bottom near the capsule. The spooler held almost 20,000 feet of ⅜-inch-diameter Kevlar line with a double braid polyester jacket. This was plenty strong enough to lift *Liberty Bell 7* to the surface. The next task was to connect the end of the lift line to the capsule using a modified steel shackle. Finally, the ROV would lift the spooler up to the surface, unspooling the line during the ascent.

I had concerns about certain steps in the procedure but couldn't come up with a way to mitigate the risk. My biggest fear was that one of the legs of the three-point sling (attached previously) would get fouled under one of the recovery tools and rip it off, damaging the escape tower ring. I also wondered about the initial lift when the slack in the recovery line was taken out. Would ship motion yank the tower ring from the capsule during the initial lift? It was impossible to say, and we had no ability to observe the lifting process with the ROV because we needed the vehicle's deck gear to raise the capsule. It was all a gamble, and I prayed that we held good cards.

By 10:30 that morning we had completed the final connection to *Liberty Bell 7*. There was now a line going directly from the capsule to our ship. Several hours later, the ROV and spooler were at the surface, and before long, we had stripped the line from the spooler and fed it into our deck gear. It was about to get interesting. Steve Wright motioned to Richard Parks, manning the controls for the traction winch, shoving his thumb into the air and screaming, "Start taking in." Dressed in a life jacket, hard hat, and heavy work gloves, I paced the deck like a nervous parent waiting for my child to be born. My mouth was dry from too many cigarettes, and the copious amounts of coffee I'd consumed over the past few hours had amped me up.

A little after 8:00 p.m., I asked Steve, "How much line have we taken in?"

"About 2,000 feet," he replied.

The deck gear roared in the background, a combination of several diesel engines running the traction winch and ROV system. Slowly, the traction winch rotated, hauling in the small lift line and wrapping the excess on the storage reel. Sometime before midnight, everyone on deck gasped as the recovery stopped. *What's wrong now?* I wondered. I pressed one of my steel-toed work boots on the recovery line and it felt good. Still some tension there. Steve leaned over to me, yelling, "We need to replace the alternator on the hydraulic power unit. It should not take long!" A lump filled my throat. *What next?*

It was difficult for me to not think about what was going on far underneath us. The ship was heaving up and down, only not bad because the seas were moderate. But looking at the line surging, I could only imagine the spacecraft bouncing up and down in the darkness. Would everything hold together? It was impossible to say.

I felt the lift line with my gloved fingers. Sometimes it felt like something was there; other times not. I couldn't tell. My question was answered at 2:15 the next morning as we observed the top of *Liberty Bell 7* breaking through the surface amid chunks of corrosion falling off the recovery compartment and splashing into the water. We still had it. (The August 2, 1999, issue of *Newsweek* quotes me as saying, "It was like an apparition, a ghost rising from the depths." I don't remember any of it.) As more of the spacecraft revealed herself, I saw the original recovery line dangling down the side as we slowly lifted the capsule farther out of the water. I yelled to Steve, "Let some of the water drain out of it!" He nodded in agreement as a small river of seawater and corrosion cascaded out of the open exit hatch.

Finally, the moment of truth arose. With the spacecraft gently swinging, Steve and I steadied the titanium cone as we guided it onto the rusty deck. To me, it felt as though our ship was the only point of light in the entire Atlantic Ocean. Outside of this small circle of life and light, nothing else existed. With incredible gentleness, Richard Parks manipulated the crane until the capsule was on deck, stained with mud from the bottom, seawater dripping down the corrugated shingles, and the braided stainless steel recovery line still attached to the Dacron recovery loop. Instinctively I turned around and, seeing all the people on deck for the first time, raised my arms in triumph as the throng clapped and cheered. Jim Lewis, the pilot of *Hunt Club 1*, walked up as I removed the 1961 recovery wire from the capsule and handed it to him. "Here's the other end of your helicopter," I said. Jim chuckled, cherishing the moment. It was over.

Looking back on the *Liberty Bell 7* project twenty-plus years later, I still find it amazing that we were successful at all, given the numerous things that could (and did) go wrong. We managed to succeed due to the hard work and experience of our team.

Obviously, it was within our technical capabilities because we were successful, but we could easily have failed. Some of the obstacles were within our control, but many were not. All it would have taken was an ocean swell to make the OE 6000 sonar pitch at the wrong time and we may have missed detecting the sunken spacecraft. The towering sand waves on the bottom of the Blake Basin were not any help. We had no control over where the spacecraft landed on the bottom; had it been in between the crests of the terrain, it might still be down there, yet to be found.

If the operation were done today, it would be conducted with more advanced technology. The turn times dictated by the 4,800 meters of seawater using a towed sonar were impossible to eliminate. It took us anywhere from eight to twelve hours to maneuver the ship from the end of line to the start of line on each search pass. These days, using an autonomous underwater vehicle, like the Hugin 6000 I procured for the Navy's undersea operations program, would do away with all that wasted time. However, there is always a price to be paid—in this case, the finicky nature of such equipment. Of course, we had our share of problems with the OE 6000, too, mostly self-inflicted. There is no doubt that an AUV would have given us much higher-resolution images, allowing us to discern the spacecraft more easily against the rugged terrain.

The salvage procedures we used in 1999 are still used today. There are only so many ways to raise an object of any size from such depths, and the *Liberty Bell 7* recovery is still the deepest commercial salvage operation on record, according to Guinness.

When Jeff Bezos recovered mangled wreckage of the F-1 engines from the Apollo 11 mission in 2013, he was showered with public praise from NASA Headquarters. Fourteen years earlier, after we located and raised an intact and pristine Mercury spacecraft from comparable water depths, NASA was strangely silent. In fact, I don't recall HQ saying anything about the event. Why? I think it was because NASA just wanted to forget the whole thing because it was bad optics to publicly recall the day they lost their spacecraft and almost drowned their astronaut.

The project didn't exactly have a positive effect on my career at Oceaneering. In fact, I was privately reprimanded by my boss for "neglecting my Navy duties." This after bringing in over $3 million worth of business to my employer and having them linked to a very positive international story. I left Oceaneering soon after that.

Ultimately, I think I was just the right person to come along at the right time. I personally remembered Grissom's flight from my youth, and eventually developed the technical expertise to formulate a feasible plan to find and recover *Liberty Bell 7*.

I was disappointed that the Grissom family declined to be interviewed for the documentary film the Discovery Channel produced about the expedition in December of 1999. All I know is that there were discussions between them and Mike Quattrone, the head of Discovery at the time, but for some reason, the Grissom family chose not to participate. I feel they would have added much perspective to the film. For whatever reason, Betty Grissom publicly denounced the project, according to her statements to the Associated Press. Why? I will never know. I felt that what we were doing was a good thing.

11

2000: Heavy Cruiser CA-35

Japanese submarine slammed two torpedoes into our side . . . We was comin' back from the island of Tinian to Leyte. Just delivered the bomb—the Hiroshima bomb.

Eleven hundred men went into the water. Vessel went down in twelve minutes. Didn't see the first shark for about a half an hour. Tiger. Thirteen-footer. . . .

What we didn't know was our bomb mission had been so secret, no distress signal had been sent. . . . They didn't even list us overdue for a week. Very first light . . . sharks come cruisin'.

Y'know the thing about a shark, he's got . . . lifeless eyes, black eyes, like a doll's eye. When he comes at ya, doesn't seem to be livin' . . . until he bites ya. And those black eyes roll over white, and then . . . oh, you hear that terrible high-pitch screamin', the ocean turns red, and spite of all the poundin' and the hollerin', they all come in and they . . . rip you to pieces. . . .

Noon the fifth day . . . a Lockheed Ventura saw us. He swung in low . . . and three hours later a big fat PBY comes down and starts to pick us up. Y'know, that was the time I was most frightened, waitin' for my turn. I'll never put on a life jacket again. So, eleven hundred

men went into the water; three hundred sixteen men come out, the
sharks took the rest, June the 29th, 1945.
Anyway . . . we delivered the bomb.
SAM QUINT

Such was the chilling monologue delivered by actor Robert Shaw as Sam
Quint, the shark hunter, in the blockbuster motion picture *Jaws*. However,
unlike most Hollywood versions of factual events, Quint's description of the
sinking of the heavy cruiser USS *Indianapolis* (CA-35) was pretty accurate.

The story of the *Indianapolis* is as multilayered as the depths of the ocean
where the ship sank just after midnight on July 30, 1945. The *Indy*, a favorite
of President Roosevelt, was tasked with a historic mission: deliver key atomic
bomb components to Tinian atoll in the Pacific Ocean and do it in the short-
est possible time, because every day the bombs were not in the belly of a B-29
lengthened the war by the same amount. And deliver the *Indy* did, setting a
still-standing nautical speed record from San Francisco to Pearl Harbor (an
average speed of 29 knots). But it was after dropping off the "gadget" that
things went terribly wrong.

Following another transit to Guam, the *Indy* set off once more from Apra
Harbor, this time to support the planned invasion of Japan. Commanding the
cruiser was US Navy Capt. Charles B. McVay III, one of the service's bright-
est officers. A chain of events would eventually result in his death, though not
at the hands of the Japanese, but by his own service revolver twenty-three
years later.

The number of dominoes that must be toppled to explain the loss of the
Indianapolis almost make it seem as though the ship's demise was decreed
by fate itself: McVay was convinced into thinking that there were no special hazards lurking during his trip from Guam to the Philippines, but in
fact a ship had been torpedoed and sunk on his planned route less than two
weeks earlier. The *Indianapolis* also just happened to be in the wrong place at
the wrong time: steaming on the well-established "Peddie" route, where Lt.
Cmdr. Mochitsura Hashimoto was waiting with his ten-month-old submarine cruiser, the *I-58*. McVay's ship was also doomed by the heavens, in that
the Moon was in just the right place to backlight the *Indy* as a perfect target
during a night attack. It only took two torpedoes (six were fired) and twelve

minutes for the *I-58*'s thirty-six-year-old captain to send McVay's ship to the bottom in flames, along with over 400 of her complement of 1,196 men.

In fact, if not for a malfunctioning whip antenna on a Lockheed Ventura, the entire ship's crew might have perished, with the *Indy* doomed into oblivion as a ship swallowed up by the Pacific's version of the Bermuda Triangle. The ship was even symbolically obliterated from existence by its own Navy when one Lt. William Green decided to delete the *Indianapolis* from the plotting board in the Philippines because he assumed it had arrived as scheduled. Fortunately, Navy pilot Chuck Gwinn was trying to inspect this antenna when he saw a brown smudge against the cobalt blue of the Philippine Sea—an oily beaded string of death and dying, all that remained of Roosevelt's ship of state.

By the time the 316 half-dead sailors were rescued, fingers were already starting to point. And they weren't pointed at the people responsible for the Navy's worst disaster at sea; they were directed at McVay. The Navy was so determined to crucify McVay that they even brought in to testify the man who sank his ship: Hashimoto. While the Imperial Japanese Navy failed to kill McVay and all his shipmates, the US Navy succeeded in murdering his soul, by convicting him of "hazarding his ship by failing to zigzag," during a general court-martial in December of 1945. This was in spite of Hashimoto's clear testimony that it would have made no difference: he could have sunk the *Indianapolis* whether it zigzagged or not. Over the decades, accusing letters from the families of his dead shipmates took their toll, and on a cold November morning in 1968, Charles Butler McVay picked up his .38 caliber Navy-issue revolver, lay down in the front yard of his Connecticut home, pressed the muzzle against his head, and pulled the trigger. Thus ended another chapter in the history of the USS *Indianapolis*.

What I knew about the *Indianapolis* after the *Liberty Bell 7* expedition in 1999 was what I had read in Dan Kurzman's book *Fatal Voyage*. That was it. Following the successful recovery of Grissom's spacecraft, Mike Quattrone and I discussed what to do next. We both knew that Dr. Robert Ballard (known for the discovery of the wreck of the RMS *Titanic*, among others) was planning to search for the *Indy*. Why not beat him to it? Consequently, rather than taking a well-deserved rest after *Liberty Bell 7*, I immediately dove into trying to figure out where the *Indianapolis* sank. The problem was that no one really knew.

Starting in September of 1999, I began collecting everything I could find about the loss of the *Indianapolis*. During numerous trips to the National

Archives, I collected up all the logbooks of every rescue ship, as well as the findings of two separate inquiries into the disaster, one held in Guam and the other in Palau, immediately after the loss of the ship. The Palau investigation was not even declassified until almost twenty years after the sinking. But both held valuable information.

Over a three-month period, I plotted the dead reckoning course of every rescue ship, identifying what was found, such as bodies, life rafts, wreckage, oil slicks, and so forth. In the process, I created a database of every significant piece of debris located that was time-based, in that all the surface targets could be correlated from the sinking time. It was an immense undertaking. Because I did not have the ability at the time to do computer-based navigational plotting, all the ship tracks and debris sightings were laboriously plotted on a large chart on my office wall. The chart was 4 feet high and 8 feet wide, and from it I could visualize where everything was going, and hopefully guess where the ship was when it sank.

I had a lot of problems with the Navy's "official" sinking location because no one knew where it came from. Did McVay have a small slip of paper in his pants when the ship sank? Was it what was estimated by navigator John Janney when the ship was attacked? Or did it come from the USS *Ringness*'s navigator after McVay was rescued? It had no pedigree and that was a big problem. But it was all we had so we had to go with it and see if the evidence supported it. I also knew at the time that the *Indianapolis* had sighted a landing ship tank during the transit from Guam. But not knowing which LST it was made the information worthless. It would be many years before this mystery was solved by Dr. Richard Hulver, making it possible to find the ship.

Other clues ended up as dead ends. One retired Navy sailor told me that while on an LST, they had "seen the flash" of the exploding torpedoes. This was valuable information because it could triangulate the sinking location. Unfortunately, after I had his ship's logbook pulled, it turned out that they were nowhere near the *Indianapolis* when it was attacked. Just more scuttlebutt. Another story was that the pilot of a C-54 transport aircraft flying from Leyte to Guam during the night of the sinking saw what he thought was "a huge naval battle" below his aircraft. The time of his sighting supposedly corresponded with the time of the attack, and upon arrival in Guam, he notified Naval Intelligence about what he had seen. But his son was unable to find his father's flight log and, hence, it was only a tantalizing story. However, a real clue was an image taken by a crew member of USS *LST-779*, which was eventually identified by Hulver as the ship sighted by the *Indy* during the transit from Guam. A tiny, faded photograph was taken by one Floyd Lambertson, and his name was on the list of crew members during the ship's commissioning. That one picture, probably the last one taken of the cruiser, would be a significant data point before the Paul Allen expedition. However, in 1999, I knew nothing about it. I was starting from scratch because no one had ever searched for the *Indianapolis*.

One of the people I reached out to was Harlan Twible, an ensign and US Naval Academy graduate who was assigned to sky aft when the ship was torpedoed. Harlan, a wonderful man, provided insight into how the *Indianapolis* was run and how they navigated in 1945. He was on watch when the *Indy* sank and supplied important information concerning a critical navigational fix at 8:00 p.m. on the day of the sinking:

As you may be aware, I was on watch the night we were sunk. I had the 2000–2400 watch in sky aft, a long way from the bridge or the wings where any celestial sightings would have been made. I have testified that the weather was so bad on the night that we were sunk that I had to leave my watch station to ensure that the weapons were properly manned. I gave this testimony at the Board of Inquiry a couple of weeks after the sinking.

Could a celestial sighting have been made at 2000 hours? There is no way for me to say, categorically, that one wasn't made. All that I can say is that I do not believe a sighting was possible. (Having been drilled in navigation at the Academy, I had a pretty good idea of its limitations because of weather and other factors.) This meant that any entry in the log probably was done on a DR [dead reckoning] basis. The navigator wouldn't have been one to say that he used celestial navigation when he didn't. I can also say that we didn't have Loran on the ship. If it had been put on at Mare Island, we would have all been driven by curiosity to see it and work with it. We did have it on the CL119 (Juneau) when it was commissioned in 1946 but even then it was of limited use. Of course, the DGPS [differential global positioning system] that we have today was beyond our dreams in those days.

Sorry that I can't give you absolutely definitive answers to your questions. They are beyond my ability to do so. However, taking everything into consideration, I can state that I believe DR was probably what we were using at the time of the sinking.

Harlan Twible's opinion carried considerable weight because he was one of the few surviving line officers on watch when the *Indianapolis* sank and, as stated, was educated in celestial navigation at the US Naval Academy. In further correspondence he stated:

> Knowing the captain as well as I got to know him during the years after the sinking but before his death, I respected his ability as a sailor. I also had a great deal of respect for the navigator, Cmdr. Janney. If we took a fix at 1200 and it showed that we were off course to any degree, the orders to the bridge would have been to get us back on course. However, I cannot have any way of confirming that. While the seas that day were rolling, I do not believe that they would have caused the ship to veer too much off course. As for the zigzagging, our termination of the zigzagging would have left us short of where we would have been had we not zigzagged. It would not have left us off of track. Finally, the *Indianapolis* was a well-designed ship and handled well. There was nothing in the ship's design that would have required us to carry any rudder to steer a straight course.

In later conversations, Harlan commented on when the last celestial sighting from the *Indianapolis* could have been made:

> The last celestial fix by the navigator probably occurred on the 28th if at all. We pulled out of Guam around 9:00 a.m. on the 28th. We were sunk about 39 hours later. At roughly 16 knots we would have traveled only about 175 miles at 2000 (i.e., on the 29th). We could have had a pretty accurate DR fix in that short a period of time. Janney's orders could still have been to get celestial fixes on schedule. Again, I am guessing.

Starting in late 1999, my initial focus was to study the deck logs of all the rescue ships (about a dozen) and extract any information regarding where survivors were rescued as well as where and when any floating debris was observed by surface ships. All this information was entered into an Excel database, along with related data such as the ship making the observation, date, and time of the sighting. The locations were then plotted and corrected for surface drift, considering the elapsed time, current speed and direction, and weather conditions. In addition, the various objects seen by the rescue ships (i.e., wreckage, life rafts, bodies, and survivors) were mathematically modeled and, in conjunction with historical weather and surface current data, used in a computer simulation (i.e., a computer-aided simulation program by Wagner and Associates in Norfolk, Virginia) to confirm the projected sinking location of the *Indianapolis*. The program is identical to that used by the United States Coast Guard to determine the loss location of aircraft and ships based on where flotsam and survivors are found.

I also considered the accuracy of the celestial navigation in use at the time as well as information obtained from the few officers that survived the disaster who are still living. This was considered in light of the established accuracy of 1945-era navigation and dead reckoning techniques, as well as the planned route of the ship from Guam to Leyte.

The result was that a 540-square-nautical-mile search area was established that had a very high probability of containing the target, considering the navigational accuracy involved, circumstances of the sinking, and surface current and weather conditions during the loss. Of course, as later events would prove, the Navy's official sinking location was off by more than 30 nautical miles. But I didn't know that at the time because Dr. Hulver had not yet identified the LST sighted by the *Indy*. That single data point eventually identified the actual sinking location. Other than the one Navy position and the survivor locations, the only other fix was from Capt. Hashimoto of the *I-58*. However, his attack location was more than 20 miles to the north; a big difference to say the least. I also worked with Blue Water Recoveries of Midhurst, England, as David Mearns, an old Oceaneering International (OII) coworker, had found several WWII wrecks with the Ocean Explorer (OE) 6000 and I valued his input on the operation. More importantly, he had real technical data on the accuracy of WWII sinking locations, having found ships and compared the recorded versus the actual loss points. All I knew about was FPS-16 tracking radars and Project Mercury.

Once Mike Quattrone had committed to the project (after funding my study), the question was who to hire to do the search. I wanted to use OII's

OE 6000 and Magellan 725 remotely operated vehicle (ROV) because they were systems I knew were up to the task. I had already discussed the project with Ron Schmidt, the project manager, and he was confident they could search in the rough terrain expected. The problem was that Discovery refused to work with OII. Instead, they forced me to put out requests for quotes from anyone else with the required gear. It was a short list. I really didn't understand Discovery's mindset. The system used on *Liberty Bell 7* had found more shipwrecks in deep water than any other sonar in the world. I can only assume that they didn't like how OII treated them during contract negotiations. In other words, they were used to being coddled, and OII treated them like any other client, which they didn't like. I don't think OII liked working with Discovery either because the profit margins on these expeditions would not be great and there was little prospect of long-term follow-on work. In addition, any mishaps or problems on the job would be documented by film crews and broadcast for the world to see. I can understand their reluctance given the way they were represented by previous internet "reporters." So, OII was out.

I ended up retaining Williamson and Associates, a well-known survey company based in Seattle, Washington. They were a small company and owned the SM-30 towed sonar, a low frequency system like the OE 6000. The problem was that all they did was survey operations; they didn't own a deep-water ROV system. Consequently, we would be forced to conduct an initial survey and, if we saw anything that looked like the *Indianapolis*, gamble that it was the wreck while later hiring an expensive ROV system to confirm the discovery. I didn't like it. That was because with the OE 6000 and Magellan 725, you could immediately ground truth anything you found by diving the ROV. Then you could continue searching if needed. But Discovery ignored my concerns and, like it or not, I had to go to the Philippine Sea with only a towfish on board and no ROV.

After arriving in Guam, the first thing I did was explore the island because I had never been there before and had a couple of days before the ship and Williamson's team showed up. I drove from one end of the island to the other, which didn't take long, noting several bunkers left over from the war. I also visited the local museum, which had some information on the *Indy* because Apra Harbor, in Guam, was the ship's last port of call.

Once the MV *June T*, our support ship, showed up at the commercial dock, it was all business as Williamson loaded their gear onto the ship and ran the deck cables. I knew about this ship because it had supported an earlier Earhart search expedition working out of Majuro atoll. Unfortunately, as we later found out, along with the foodstuffs loaded there were thousands of roaches and every kind of Pacific insect you could imagine. The ship was owned by International Bridge and Construction and normally transported building supplies to all the nearby Pacific islands. It was a typical workboat— not exactly new, but it was based in Guam and available so there were not any transit costs to pay, which helped because we were tight on money.

Like the OE 6000, the SM-30 sonar used a two-body tow system. In other words, the end of the long, armored tow cable was attached to a depressor, which helped dampen out ship motion. About 100 meters behind the depressor, on a soft umbilical, was the search vehicle. It was fitted with two sonar transducers as well as a sub-bottom profile, which could blast low frequency sound into the sediment below the vehicle. It was a proven system but had no way to image what was found, other than to display computer-processed sonar targets.

Following equipment mobilization, the support ship *June T*, with myself, a small film crew from Partisan Pictures, and the search team, departed Apra Harbor on June 1 for the work area, approximately a three-day steam from Guam.

We arrived on location in the early morning hours of June 4 and launched Williamson's wide-swath SM-30 sonar into the Philippine Sea. Initial search operations went well as the sonar examined the location using the 5-kilometer swath during 30-nautical-mile track lines and had completed four such search lines by June 8.

In the year 2000, information on the terrain for this area was essentially nonexistent. There was no Google Earth, or even much of an internet at the time. We had standard nautical charts and one very large-scale bathymetry map with contour lines spaced wide apart. Even so, we all knew the bottom terrain might be bad. But how bad? Impossible to say. Would it be so bad that we couldn't even tow a sonar? We didn't know. Autonomous underwater vehicles were not yet available for commercial use. Everyone still used the towed sonars for deepwater search operations.

I vividly remember the first time we arrived at the Navy's attack position. I was lying back with Don Campbell, the film's producer, on the bow as we neared the spot. For anyone who has never been far out to sea, away from the smog and light pollution, the sky is truly impressive. On a darkened ship, after about thirty minutes your eyes adjust to the low light. When you look up, you can see a dark sky sprinkled with millions of stars. It makes you understand how insignificant we humans are in the solar system, our galaxy, and the universe. It is an emotional experience when you see it for the first time. I remember saying to Don, "I guess this is not such a bad place to die..." I said it because that's what had happened. Hundreds of young men died seeing what I was seeing.

Our initial impressions of the location indicated that the bottom was quite irregular, with vertical reliefs of over 3,000 feet in some cases, when looked at from the direction of west to east. As a result, we made sure we had sufficient overlap between the 5-kilometer search lines to allow for adequate coverage of the area. As we surveyed the bottom, we also noted that the largest percentage of promising sonar targets seemed to be centered in the general area where the *Indianapolis* was reported to have been sunk, some of them being hard returns amid bottom geology with others representing themselves as scattered point targets. Based on what we were seeing, I decided to image certain areas of interest using the sonar in the 2-kilometer swath, which had considerably higher resolution. The *June T* worked well while towing the sonar at low speeds in the westerly and easterly directions, and the weather was good with light seas and winds. However, soon after we started these high-resolution runs, the deck-mounted cable reel began having intermittent problems, making it difficult to control the depth of the sonar and guarantee that we could clear certain bottom terrain in time to prevent impact.

Unlike the OE 6000's, Williamson's cable reel did not use a traction winch to haul in the cable behind the ship (which would lessen the strain on the main storage reel). Consequently, when we had problems with the reel, it affected our ability to pay out and haul in cable to clear the terrain. After we dealt with this problem for several hours, the cable reel failed completely. We were steaming around the Philippine Sea with thousands of meters of cable dangling behind the ship and no way to bring it in. We couldn't stop, because the search vehicle would land on the bottom amid the boulders and cliffs. But eventually we would have to stop and figure out a solution because the ship had only so much fuel. Finally, after jury-rigging a temporary fix using the ship's hydraulic pump, we got the cable back in. Unfortunately, the main hydraulic motor was shot, and the deck was now covered in slippery hydraulic oil.

So, we made the tough decision to head to Palau, the nearest port, to see if we could get new parts and fix the thing. It was more than a day and a half away. More wasted time. Soon we would be out of everything—time, parts, good weather, and money.

By June 11 we were in transit to Palau, where no parts would be available. But at least we would be somewhere they could be shipped to.

While en route to Palau, one of the *June T*'s assorted insects bit me on my right foot. That was sometime after I spewed vomit over the starboard rail after eating dinner. By morning, the wound was very red and swollen. It was infected. I didn't know what it was, but I suspected a spider. I was limping all during the transit to Palau. Fortunately, the captain had just replenished the medicine cabinet and I was able to smother the area with antibiotics. In the Pacific, such things, especially at sea and if not treated, could cause the loss of a foot. There was no hospital, no clinic, no doctor, just the old captain from New Zealand who showed us pictures of himself in front of his biplane. Fortunately, the wound healed, and I kept the foot.

We didn't leave the island until June 14, and after we made some attempts to search in the area, the winch motor failed again. This time, Palau would not do; we were headed back to Guam. Discovery was not happy. But what could I say? I had wanted to use Oceaneering. Three-plus days later, we made it to the commercial port in Guam, once again waiting for parts. We were no closer to finding the *Indianapolis* than when we left. We had some sonar and sub-bottom targets, but without any way to determine what they really were, it was a crapshoot at best.

It would not be until June 22 that we got underway for the search area again. Three days later, we were finally doing high-resolution passes on what we had for targets. It was a challenging experience. I remember one run where we almost crashed the towfish into one of the area's numerous underwater cliffs. The bottom was coming up fast as Mike Williamson, with sweat on his brow, exclaimed, "Keep taking in cable." The towfish continued to the bottom. He then added, "Increase your take-in speed!" The bottom loomed even closer.

Gripping the microphone in his hand, Mike shouted, "Bridge, sonar. Increase speed to 4 knots!" The *June T*'s engines throbbed as the SM-30's tow cable tightened and strained aft. Finally, after a few harrowing minutes, the bottom dropped away as the towfish cleared the rocky terrain by 187 meters (we normally towed at 500 meters off the bottom, in the 5-kilometer-range scale)—a close one given the water depth of 4,500 meters. Such was our experience using a towed sonar on top of the Kyushu-Palau Ridge.

We continued to inspect our best targets for over a week, hoping that one might be the *Indianapolis*. Some of them were point targets scattered throughout a basin area. Mike thought they looked like debris. But with no way to see them firsthand, it was only a guess. I liked target 1071, which had the appearance of the stern of a ship, complete with what looked like skid marks on the bottom after sliding down a slope. It had some good shadow, which suggested structure on the top. The problem was that the size measurements using the sonar processing software were dubious at best.

Donna, Williamson's sonar processing expert, was very, very good at examining targets but noncommittal on what they were. The problem I had was that as we measured the size of these acoustic images, they kept getting smaller. I knew how large the *Indy*'s wreckage might be, assuming that 60 feet of the bow were missing, as reports during the sinking suggested. Some of our best targets, scattered among the geology, were initially estimated to be the right size, and they all had decent shadows, implying height off the bottom. But as Donna continued to examine these contacts, they always seemed to get smaller. I wasn't even sure that she had ever seen a shipwreck on sonar during an operation. At the time, I hadn't either. All I knew was what torpedoes, fighter jets, and Mercury spacecraft looked like. Consequently, I relied on her judgment.

When we were done with the high-resolution runs in the 1,000-meter scale, the *June T* turned east, roaches, spiders, and all, as the Williamson team fed out cable to later haul it all back in to clean it with fresh water and get better wraps on the drum. At least the winch was still working. Getting that thing to function was the only decent thing to happen in weeks.

Phase II of the expedition would soon follow, but not until after United Airlines lost my baggage on the trip home to DC. I thought I'd never see it again. But incredibly, my duffel bag, complete with all of my research documents (too heavy to carry on) was dumped on my front lawn in Potomac, Maryland, over a week later. *Whew.*

After all the equipment breakdowns, the hot sun at 2 degrees above the equator, getting bitten on my foot, throwing up over the side, and so much rolling on the search lines, I was exhausted. After reflecting on the operation, with what I knew at the time, I thought one of the targets *had* to be the ship, given the known accuracy of WWII sinking locations. I was wrong.

In the time between the two expeditions, I flew down to Sarasota, Florida, to meet Harlan Twible, the last surviving line officer from the *Indianapolis*. He and his wife, Ann, had lunch with me as we talked about the mission and what we might find. One of the things I asked him was what it sounded like when the *Indy* went down. Harlan replied, "What I remember is something like the sound of a freight train going by. . . . A long rumbling sound."

Harlan and I liked each other, and I got the impression that he saw me as a son he would have liked to have. Later, when we were about to head to Guam, I asked him if he wanted to go along. The Discovery Channel really wanted him there. But he declined. I think there were three reasons for this. First, he had experienced several heart attacks and had no business being on a ship over a day from the nearest land. Second, he didn't like Paul Murphy, one of the four survivors scheduled to go (this was related to the USS *Indianapolis* Survivors Organization, of which Murphy was president). Finally, and I only realized this much later, Harlan was an officer and a Naval Academy graduate. The rest of the men on our team were ordinary seamen. Officers and sailors typically didn't socialize together.

Another thing that Harlan talked about was the "wonderful training" he received at the academy with respect to surviving a ship sinking. However, I doubt any amount of training could have prepared anyone for five days of hell floating in the Philippine Sea watching your shipmates go mad or get picked off by sharks.

Joining us on the second expedition were four survivors of the disaster: Woody James, L. D. Cox, Paul Murphy, and Mike Kuryla, all seamen on the ship. Watching the four survivors during the time I later spent with them was an exercise in human observation. All of them weathered the seas well and, to a man, never complained. They were simply amazed to be there. Paul Murphy appeared to be a calculating individual who presented himself as above the fray. Of course, he was the president of the Survivors Organization, for whatever that meant. L. D. Cox was simply a good old Texas boy and a joy to be with. The same went for Mike Kuryla; he was simply a good guy and happy to be there. For Woody James, however, a return to the sinking location seemed to bring up some deep-seated feelings. He would lean against the ship's rail for hours, simply staring at the sea while he smoked. I didn't know whether he was reliving the horror from 1945 or not. But I would have liked to know.

I was conflicted about having the survivors sail with us. On one hand, I respected them and understood why Discovery wanted them along to potentially see their ship for the first time since 1945. It would make for a much better film. However, having men of that age on board presented a host of other problems. It was another thing I would have to deal with. For example, we had to have a physician on board to tend to their medical needs. This was not a big deal because we found a retired military doctor in Guam who was qualified. But I was simply worried about them getting injured on board, especially if we hit bad weather, which was a definite possibility given when we were scheduled to sail and the onset of typhoon season. All of the storms started out as a low-pressure area, and as they built in strength they generally went right through our search area toward Taiwan and Japan.

The ship found by marine services contractor Phoenix International Holdings was the MV *Sea Eagle*, an aging workboat that had transited all the way from the Philippines to Guam. Unfortunately, it was what we could afford, and it was a real piece of rusted junk. The first thing that happened was that the Coast Guard inspected the ship and found numerous safety violations, mostly due to crew training and licensing. In addition, the *Sea Eagle* didn't have a sewage storage tank and instead flushed all its waste into the sea. This was not a problem once we were a certain distance from land, but in the harbor in Guam this was not permitted. As a result, the first thing the Coast Guard did was shut down all the toilets on the ship and put in place a single rent-a-john on the main deck. In the hot Guam sun, it didn't take long for the thing to stink so much that you felt like vomiting after using it. I hated those things, and it created a real logistical nightmare with the survivors. If we were to sail, I would need them on board the evening before. I could not ask such men to use a rent-a-john on deck, so I managed to get a local submarine charter company to let us use the regular restrooms on the pier reserved for their clients. It was another issue I could have done without. But there would be many more, mostly related to the Remora ROV that Phoenix supplied.

As I mentioned earlier, I wanted to use Oceaneering's equipment but was rebuffed by Discovery. The Remora ROV I thought I hired was the one used to survey the INS *Dakar*, a submarine lost in the Mediterranean Sea back in the 1960s. But the vehicle they showed up with was not the same one. It was a new system Phoenix had built, and it came complete with all of the problems of any new vehicle. The worst part was that its armored optical umbilical was one discarded by the US Navy as not suitable. Phoenix had bought it for a song. The problem was that after they loaded their equipment on board, all of the optical fibers were broken, requiring them to be re-terminated. In every case, there was too much attenuation, or loss of light down the fiber.

Their Remora system was loaded on the ship on August 4, but we didn't sail until August 9, using up five valuable days while we played with glass fibers. Such a small vehicle system should have taken only a couple of days to get on board and working. In the end, Phoenix was never able to properly terminate the optical fibers; we had to get an instructor from the local community college to do the work.

We finally left Apra Harbor at 8:00 that night as we rolled heavily in the swells outside the port with the winds at over 30 knots. I didn't want to leave in such conditions, but we had no choice; it was almost three days to the *Indy*'s location and the weather was predicted to be good when we got there.

A few days before we departed Guam, I spent some time with Woody, Mike, and L. D. Cox "from Apache, Texas . . . ," as he always liked to say. They came across as great guys as we sipped our beers in the hotel bar; all of them thrust together in a singular horrific event that happened so long ago, when they were young men, some of them teenagers. To a man, none of them could have imagined in their wildest dreams that they would be going back to the Philippine Sea, to the very location where they lost their ship. While the years had dulled some of their memories, I think that what happened to them would never leave their souls. It was part of them, and they would never really forget it.

By August 10, the *Sea Eagle* was laboring in heavy seas, managing to make only 5 to 6 knots in 40-knot winds. Spray was flying over the top of the bridge as we changed course to a more southerly heading for a better ride. But even so, at one point on the bridge, I saw the ship's inclinometer showing we were rolling through arcs of 20 degrees. That may not seem like much, but far above the deck on the bridge, it was pretty hard to stand up. As the ship chugged along, a tropical depression to the northwest strengthened into a tropical storm with winds of over 40 knots. It was about then that the ship's wind speed indicator failed. "Still taking heavy rolls," my log states.

The survivors of the *Indianapolis* were hunkered down in their racks because there was little else to do. None were seasick. There were no hot meals since it was too rough to cook. I think I lived off hard-boiled eggs for two days.

Don Campbell, who was bunking with the survivors, recounted to me their behavior while this was going on: "I would be trying to sleep, and every once in a while, one of them would crawl out of his bunk and I would hear this pitter-patter of feet as they went to the head to take a leak several times during the night." They had the prostates of old men, not the young bucks they were in 1945.

As much as one man could love another, all of us loved them, as survivors of a horrific event, as men, and above all, as shipmates.

As we neared the sinking location, the weather abated a little. No longer were we rolling so hard, and one by one, the *Indy* crew came out of their racks to hang out on deck.

On Sunday, August 13, I was jerked back into reality when I was told that the Remora's main high-voltage contactor had failed and we had no spare. So before managing to make even a single dive, at 2:30 in the morning, our bow turned to the south toward Palau. By then it didn't matter because most of the optical fibers in the umbilical had shattered once again from the vibration of the ship during the transit.

After being cleared by customs, I left the ship in Koror and checked into the Nikko hotel and started drinking. Of course the air conditioning was broken. To give you an idea of what Palau is like, the pilot who guided us through the coral reefs was the island nation's minister of transportation.

The only good thing to come out of our stay in Palau was drinking in the cave bar near the estuary. It was literally a bar built into the side of a cliff with tables inside of a cave. Free air conditioning. Years later, while I was working with Vulcan Inc., the place was closed and a Japanese luxury hotel was built on the land.

We also made a day trip to Peleliu with the four *Indy* survivors. It was an amazing journey through the fabled Rock Islands, and we even got to dive in Turtle Cove and on a sunken Japanese freighter. You could see 100 feet down and the sea was like bathwater—the best diving I had ever experienced. However, stomping around Peleliu was sobering. I didn't know nearly as much about the invasion then as I do now, but you could pick up cartridge cases still lying on the ground around Japanese bunkers. I remember a Japanese command post in particular, peppered with holes about 5 feet in diameter created by the 16-inch armor-piercing shells fired by our battleships. In fact, the *Indianapolis* was one of the ships that supported the Peleliu landing. The blockhouse was made from steel-reinforced concrete about 6 feet thick. But it did no good. Anyone in there would have been eviscerated by the attack. Also scattered around the island was a Japanese light tank on the one airfield as well as the remains of a Japanese Zero in a nearby swamp. What was tragic about the whole thing was that the only reason the US Marines invaded the place was because Gen. MacArthur wanted an airfield in a forward area as he pushed toward the Philippines. But no airfield was ever built. It makes you wonder why so many young men had to die in the first place. After stopping off at Peleliu's mall (essentially an open-air market), we hydrated and headed back to Koror.

We finally left Palau on Wednesday, August 16, and stood off the island to do a test dive with the Remora. The first thing that happened was the main video camera failed. Then the scanning sonar crapped out as well. We were not getting off to a good start. Finally, by 8:00 p.m. we had the ROV strapped down and began steaming back to where we thought the *Indy* sank.

We arrived at the search area early in the morning on August 18 to light winds and an imposing ground swell. As the Phoenix crew readied the vehicle, the *Sea Eagle*'s helmsman practiced station-keeping, where he held the ship in one position: standard on all ROV operations. Remora was finally off the deck and headed to the bottom at 4:25 in the morning as the crew attached numerous football floats to the armored umbilical. The floats helped to keep the vehicle's cable clear of the bottom and out of the way.

Only an hour later, a squall moved through the area with accompanying rain and thunder—a harbinger of things to come. The ship's propellers clawed at the churning seas, fighting to keep the ship in position in the rising breeze. We were sitting 500 meters east of target 1071, our best sonar contact from the search, which I hoped was part of the *Indy*'s stern. Unfortunately, that all ended shortly when the Remora suffered a GFI (ground fault interrupter) fault. This usually indicates a short between a high-voltage power wire and the vehicle's ground and is a safety device designed to keep personnel from being electrocuted. Remora was finally hauled to the deck during a dead vehicle recovery. This was not a big deal in decent weather, but in heavy seas it can be a nightmare.

What had happened was that a light boom came loose and smashed into the side of the vehicle's main junction box, where the electrical power is distributed to the rest of the various systems. The boom was supposed to be hydraulically controlled, but Phoenix didn't have sufficient plastic tubing to connect it into the hydraulics. So, it was tied off but then came adrift and smashed into the Lexan cover, breaking it and flooding the high-voltage wiring with seawater. It would take a long time to clean out and repair. We ended up removing the booms and mounting the high-intensity HMI lights elsewhere.

As it turned out, the GFI fault was not caused by the junction box damage. There was a fault farther up the umbilical, a more serious concern. Why? Because that meant we had to determine where the bad section was, cut off the cable, and re-terminate all the wiring and glass fibers.

As the work continued, we also learned that the solenoid driver board on the vehicle, which controls the pan and tilt of the camera, had failed, meaning that we could not move the camera underwater. As the fiasco continued, another squall line pummeled our ship as a low-pressure area moved toward us and strengthened into a tropical storm. The barometer was also falling—not a good sign. Consequently, the *Sea Eagle* departed the area for calmer seas.

We continued to monitor the weather and didn't get back to the dive location until Saturday, August 20. The US naval station in Yokosuka, Japan, was helpful in that they were willing to give us regular weather reports. The tropical storm was now Typhoon Bilis and headed straight for Taiwan.

The seas didn't lie down until the twenty-first. So far, we had wasted over a week on equipment breakdowns and bad weather and accomplished nothing. Phoenix finally got the Remora off deck in the morning but only made it down a few hundred meters before the dive was aborted due to another ground fault; a dead vehicle recovery again. There was seawater leaking into the high-voltage junction box and the port vertical thruster was leaking oil. This was followed by more work, more dives, more ground faults, more bad weather, and my boiling temper. There were about a half-dozen dive attempts, all followed by ground faults. In one case, the Remora made it down only 170 meters. The bottom was about 4,000 meters. After all this time, we were still on dive number 1. Phoenix ended up swapping out the high-voltage transformers on the vehicle.

It might seem that all our problems were due to the difficulty of working in such deep water. That is true to a point. But by the year 2000, ROV technology was mature, though there were still not a lot of 6,000-meter-rated vehicles. Today, in 2024, they are commonplace. Another factor was the many spare parts that should have been carried but were not. The whole situation could have been avoided if my sponsor had let me hire the equipment I wanted. That said, there are always problems working underwater with complex equipment. But all of our difficulties on this expedition could have been avoided with better planning on Phoenix's part.

On what was called dive number 2, we still had numerous ground faults that rippled through the vehicle, tripping the GFI, but we finally made it to the bottom at 1:40 a.m. on August 23. However, we were able to search for only a little while around the first target before the scanning sonar failed and we had to recover once more.

The week from that day to the thirty-first was filled with numerous aborted dives, another trip to Palau to pick up new parts, a second typhoon, seawater shooting in through the galley portholes, and watching our four survivors of the *Indianapolis* wander the deck like aging zombies.

We picked off the targets like clay pigeons, each one discarded like trash as they were confirmed as geological formations. We didn't find a single man-made object the whole time. The bottom around the Kyushu-Palau Ridge was an interesting area, if you wanted to look at rocks. In some places, it looked like molten lava had been squeezed out of a gigantic toothpaste tube. The one area that Mike Williamson thought was a debris field (from sub-bottom profiler data) was an abyssal plain filled with scattered boulders the size of Volkswagens.

I finally pulled the plug on the expedition at 7:00 p.m. on August 31 after calling the Discovery Channel and telling them that to stay there was a waste of time and money. We had located and identified numerous sonar targets, been beaten to hell by the weather, ripped apart and reassembled the Remora, and found nothing. With another typhoon headed right for us, there was no point in staying there and pissing away more money. It was over, at least for now.

When I returned to the States and after the dust settled, I was once again unemployed. My life had taken a nosedive since *Liberty Bell 7*, and my then wife and I were having financial problems. The long-term contract with Discovery never materialized. Mike Quattrone left Discovery to make a living as an independent producer; Sarah Hume, our producer on the *Liberty Bell 7* project, did well at Discovery, eventually rising to the level of executive vice president.

By then I was not getting any younger and could not find any work in the subsea industry. Eventually, I was reduced to digging ditches for an electrical contractor in Maryland. My wife continued to spend money with no concern for the consequences and eventually we had to refinance our home, which drove up our mortgage payment even higher. I was in dire straits.

The next year, the unidentified suspected wooden shipwreck we found during the *Liberty Bell 7* search garnered new interest among a few friends and investors. I went back to Oceaneering as a visitor and recalculated the target's position from our original sonar and navigational files. I had a position, one that Mike McDowell, holder of a long-term charter of the RV *Keldysh*, a large Russian research ship, thought a few of his supporters might be willing to dive on and contribute funding for. The primary interest in what was eventually called "Atlantic Target" was that it was lost right on the same route taken by Spanish almirantes and capitanas while they transported gold and silver back to Spain after they plundered the New World. In other words, it could be a treasure ship. Even though I was paid only for my expenses, I signed on so I could stop digging ditches for a while.

12

2001: The Deepest Shipwreck

I stared at the circular opening above my head with some concern, thinking, *Uh, I can still see daylight through that*. The aperture I was concerned about was the thick hatch in the roof of the Russian Mir I manned submersible we were just about to use to plunge 3 miles down to the bottom of the Atlantic Ocean. That's right—3 miles.

Accompanying me inside our vehicle of "inner space" was Dr. Anatoly Sagalevich, famous for taking film director James Cameron down to the sunken ocean liner *Titanic*, and Jim Sinclair, a marine archaeologist well-known for his work on the Spanish galleon *Atocha*. In addition to my two passengers, I was surrounded by miles of electrical wiring, oxygen bottles, carbon dioxide canisters, scanning sonars, video monitors, and other assorted mechanical devices, all of it needed to keep three humans alive in the crushing pressures of the deep ocean.

Before I pondered my fate further, Anatoly stood up and, gripping a small white wheel with his strong hands, locked the hatch into place. No more daylight. We were sealed in. With that, I heard some garbled words in Russian erupt from the radio and felt a gentle bump as a massive crane plucked us from the deck of our ship and lowered us into the sea. Once floating over the side, I heard a few footsteps on top of our 18-ton submersible as we were released from any connection to our support ship, the *Akademik Keldysh*. Anatoly grabbed the microphone, grumbled some more Russian, and flipped some switches, allowing water to flood the Mir's ballast tanks. We then promptly sank.

By that day in July of 2001, I had spent some twenty-five years working in the subsea field, mostly as a pilot of remotely operated vehicles, or ROVs as they are commonly known. During my last trip to that particular spot of the ocean, identified as the Blake Basin, I led an expedition to locate and recover the *Liberty Bell 7* Mercury spacecraft flown by astronaut Gus Grissom in 1961. While the flight had been a success, the explosive hatch on the capsule jettisoned prematurely, allowing the spacecraft to sink, and giving me a historic target to find and raise some thirty-eight years later. But it was what we found while searching for *Liberty Bell 7* that was the focus of our renewed efforts as we sank like a stone into the darkness of the abyss: a mysterious sonar target having all the characteristics of an old wooden sailing ship. What made it even more intriguing was the fact that our location was almost directly on the route the Spanish galleons had sailed after being loaded up with gold and silver in the New World. Both a small group of supporters and I hoped that was what we would find after viewing the target firsthand. But for all we knew, it could be one of Bill McCoy's famous rumrunners lost during a storm—in other words, a potentially worthless endeavor. For our purposes, we code-named the object "Atlantic Target."

The Mir submersible I was riding in had been used for everything from filming the *Titanic*, to inspecting the sunken Russian submarine *Kursk*, to taking wealthy tourists down to the sunken German battleship *Bismarck*. Two Mir subs existed at the time, both developed by the P.P. Shirshov Institute of Oceanology for scientific research. They were manned submersibles, each of them able to carry one pilot and two passengers to the bottom in waters as deep as 6,000 meters, or roughly 20,000 feet (that capability allowed them to reach 98 percent of the ocean floor). Both vehicles operated from the *Keldysh*, which at 6,500 tons is the largest oceanographic vessel in the world. On that ship, you can do anything from observing a Russian scientist studying a deep ocean species of fish, to watching a movie, to getting soundly trounced in a game of volleyball (the Russians are pretty good at it). And if you look very closely at the *Keldysh*'s smokestack, you can see where the bright red hammer and sickle insignia has been covered in white paint; an example of Communism wiped out in more ways than one.

Before we dove, I shared some of the side-scan sonar data with Anatoly, describing the towering sand waves known to be in the area. He was unimpressed. We later had a dive meeting at 8:30 in the morning before we launched the Mirs. Most of the Mir support team huddled in the laboratory, basically a cramped workshop used to keep the subs operating, as Anatoly reviewed our objectives. "Today we dive to try to locate a target . . . It is unknown . . . But who knows? Maybe a Spanish galleon . . . maybe not. We will find out." He ordered all the "divers" to be back in the Mir lab at 9:30 and sent us all off to get ready.

The Mir's exterior size was deceiving. Most of the vehicle's volume was needed for two purposes: to keep three occupants alive and to transport them across the bottom in deep water. The three of us would be crammed into a 7-foot-diameter nickel–steel sphere mounted at the sub's bow. The rest of the sub's length was filled with batteries, ballast tanks, propulsion units, hydraulic pumps, and electrical systems. Generally speaking, diving in such a research submersible is fairly safe; to my knowledge, there has not been a fatality in a similar vehicle since 1973. But the reality was that if we somehow got stuck on the bottom and could not be unfouled by our sister sub, we were dead. I knew that there was no way another vehicle, one capable of reaching such depths, could be mobilized in time to rescue us. After a few days, our oxygen and carbon dioxide absorbent would run out, the batteries would die, and we would be nothing more than a lifeless hulk on the seafloor. Eventually we would be found, our bodies removed, and the sub cleaned out and refurbished. However, I had faith in the technology created by the losers of the Cold War.

We would also be working in an area that, like most of the deep ocean, was generally unexplored. In fact, we would be the first people to visit the location in person. More humans have been into space than have dived to the depths we were about to visit—and hopefully return from as well.

In the days since sailing from Bermuda, I came to respect the Russians, Anatoly in particular. He and his team were very resourceful. In an age where he saw his whole government collapse, Anatoly found a way to keep his organization running. The man was a burly individual with a balding head and was no doubt idolized by his people. When I first came on board the *Keldysh*,

Anatoly was cordial, but not overly friendly. I judged him to be someone from whom you had to *earn* respect. During our transit from Bermuda to the dive location, we all watched the Discovery Channel film depicting my expedition to recover *Liberty Bell 7*. After that, Anatoly seemed to accept me as a comrade underwater explorer. I suppose he figured that if I could find and recover a tiny Mercury capsule from such deep water, I must know what I'm doing.

Preparing for a deep dive is a process, much of it physical, the rest mental. The day before you know you're diving, you don't eat or drink much, mostly due to the difficulty of relieving bodily fluids and getting rid of solid waste in the sub. It's not that it can't be done, it's just that you want to avoid having to do it. It is an easy process to urinate into a special container during a dive if you must, but defecating is another matter. Imagine using a toilet in a stall that is also occupied by two other people trying not to watch you. That's what it would be like.

After our morning briefing, I returned to my cabin and put on the bottoms of a pair of thermal underwear, a T-shirt, thick socks, and a blue Nomex jumpsuit supplied by the ship. If anyone had told me ten years earlier that I'd be diving in a Russian research submarine wearing a suit with a patch embroidered with the letters "CCCP" (the acronym for the Cyrillic alphabet spelling of Soyuz Sovetskich Sotsialisticheskich Respublik), I would have told them they were nuts. Yet here I was.

Promptly at 9:30 a.m., I made my way down to the Mir lab and signed the dive log so that a record was made of who was diving in which submarine (in case we didn't return, I assume). I gave my small bag to the aging lady who later placed it into the appropriate sub. I walked out on deck and climbed a narrow aluminum ladder up the side of Mir I, which was still being prepped by the crew. Once I reached the top of the sub, a few people nearby started clapping, and I gave a small wave while removing my shoes and handing them to a crewman, who stuffed them into a small toolbox for our return. There is almost a tradition to it all, the simple act of diving in the Mirs.

Inside it was very warm, probably about 90 degrees Fahrenheit with a humidity of 100 percent. I crawled onto the left side couch, pulled down my jumpsuit, and stripped the clothing off my chest. I stuffed my small bag toward my feet and waited. Anatoly and Jim soon followed, the hatch was sealed, and we were off. I didn't know Jim Sinclair well at all, but the husky, bearded

man was well-known in treasure hunting circles and one of the few marine ar-
chaeologists willing to work with commercial salvors. Jim also took the time
to point out two toggle switches on the right side on the forward control panel.
If for some reason Anatoly became incapacitated, pushing those two switches
up would return us to the surface. Good information to have.

We sank further still into the abyss. Inside the Mir, it was quiet except for
the sounds of our breathing, occasional communications, and the humming
of the electrical systems. As we passed 1,000 feet, we lost all our ambient light
and the cobalt blue of the Bahamian waters turned to a deep black. It also
started getting cold so I pulled my clothing back on and slipped on a second
pair of socks. There was really nothing to do at that point, and I tried to relax
on the thin padding of the observer's couch.

Two hours after leaving the warm sunlit environment of the surface, we
neared the bottom of the Blake Basin. By now, the outside water temperature
had crashed to a frigid 36 degrees Fahrenheit, droplets of water were dripping off the sub's interior, and the pressure on every square inch of our pressure sphere was about twice the weight of the average passenger car—over
7,000 pounds. As our long-range sonar started to detect the nearing seafloor,
Anatoly flipped some additional switches and began pumping water out of
our ballast tanks to slow our descent. Crashing into the bottom would be bad.

The seafloor came up to greet us. I saw the side of a slope of mud pass us as
we plummeted down. *Damn, these hills are tall!* I thought. The bottom muck
swallowed our submarine, and we finally came to a halt. We were now at a
depth of 16,374 feet beneath the surface. I peered out through the tiny viewport but could see nothing but a swirl of bottom sediment and the side of a
very steep escarpment. Anatoly was now impressed with the bottom terrain.
"Hills are very big," he mumbled. He continued to expel water from the Mir
as we slowly began to rise off the bottom and then head west in search of our
long-lost sunken ship. The swift currents in the area had shoved us far to the
east of where we wanted to be; strong currents in such deep water were unusual, to say the least. Now we had a column of water above our tiny sub equal
to the height of twenty-eight Washington Monuments.

One thing I noticed was the difference between observing the bottom on
a video monitor and viewing it firsthand. With a remote vehicle, all you get
is a two-dimensional image, one viewed while holding a cup of coffee in your

hand and sitting in relative comfort in an air-conditioned control van. You know that if anything happens to the vehicle, all you have to do is throw up your hands and go back to bed while the ROV is recovered. In the Mir, I saw with my own eyes the steep edge of a mud wall disappearing into the blackest black you can imagine. I felt as though I was hovering above a bottomless pit of darkness, held up only by the couch I clung to. When the sub's skids scraped the bottom, I saw the sediment swirl across my viewport and felt my forward motion grind to a halt. I smelled the stale air inside of our sub, felt the ice cold just outside, and heard the throbbing sounds of our electric motors. It was as if I was part of it all, as opposed to a detached observer. It was real, not imagined.

We drove across the bottom, right through my best estimate of the sunken ship's location. Nothing. Thirty minutes after we hit the water, Mir II also dove and joined us on the seafloor, even farther to the east. We ordered them to work in the area north of the target location while we zigzagged around the southern area.

As Anatoly flew the Mir, he skipped over the crests of the massive sand waves with ease, except for times when he would nick a hill with one of the skids. When that happened, it was like running into a wall of jelly as the submarine suddenly lurched to one side and ground to a halt. I was reasonably comfortable in my steel prison, except I had to keep pushing myself up the couch. This was because as Anatoly drove forward, the vehicle was pitched up at about 10 degrees, causing me to continually slide toward the rear of the pressure sphere. I then would use my toes to push myself back up, and in the process ended up with my feet soaked with condensation. My toes froze. Also, it was awkward to look out of my viewport. I had to lie on my right side, pressing my ribcage into the hard couch, and hold my face against the freezing cold steel of the area surrounding the viewport.

All of a sudden there appeared a suspended steel cable, directly across our path. I gestured to Anatoly, and without saying a word, he nailed the Mir's electric throttles to their stops and skated the sub over the potentially deadly obstacle. Dancing with communications cables was one excellent way to get stuck on the bottom. Now I remembered. When I was doing research on the *Liberty Bell 7* project, I examined an AT&T cable chart that showed the location of all of the active and decommissioned communications cables off the coast of Florida. There was one telephone cable near the north of our search area, and we had just barely missed getting snagged on it.

After six hours on the bottom, we still had found nothing, and I was beginning to have serious doubts about whether anything was there at all. But I remembered that thirty minutes before we found *Liberty Bell 7*, I had had similar feelings of failure.

It was about then that the underwater telephone crackled to life. "Mir I, this is Mir II. We have found a wooden shipwreck!" Our tired, cold faces erupted into smiles all around as Anatoly and I started plotting a course to find the other Mir. More details were transmitted as Victor, the pilot of Mir II, said, "We can see the copper sheathing on the hull . . . and a large pile of coins!" That took us by surprise. Could we have indeed found a treasure ship?

Jim and I conferred. "That's amazing, isn't it, finding coins this early," I remarked.

Jim rolled over to face me, replying, "Usually when they find coins so soon it means there are a lot more there."

But try as we might, we could not find Mir II. We got close enough to hear their propulsion units churning away, but the towering sand dunes kept them hidden from view. One problem was that our navigation system didn't work very well in such notorious bottom terrain. The sound from our navigational beacons was being blocked by the massive hills.

Frustrated and exhausted, Anatoly finally said, "Battery power is low. I'm sorry, but we must surface." With that, we all resigned ourselves to the inevitable and Anatoly began pumping water out to return to the surface, the whining motor sounding like a tortured animal. We had to leave Mir II and *my* shipwreck on the bottom for now. As our depleted batteries squirted out a little more juice, I heard the slow whine of the hydraulic motor as it pushed water out against the enormous pressure at depth. I had been told earlier that you really got cold on the return to the surface and they were right. I was shivering as I wiggled my body into a thick thermal suit. Even that wasn't enough, so I also covered my hands with woolen gloves, put on a third pair of socks, and covered my head with a watchman's cap. With nothing else to do except listen to water being expelled from our submarine, I curled up into a fetal position and drifted off into a semiconscious state. I also had a pounding headache from the high concentration of carbon dioxide in the atmosphere. By then we had been confined inside of the Mir for almost thirteen hours.

When I awoke from my partial sleep, Anatoly seemed to glance at me with smiling approval. Had I passed some sort of test because I was relaxed enough to sleep while underwater? Had I proved that I was bold enough to dive 3 miles down in his submarine? We were nearing the surface and starting to roll around a little in the swells. Now I really had to urinate and couldn't wait to get out of the Mir.

Within minutes, our wallowing submarine was hauled out of the dark seas onto the brightly lit deck of the *Keldysh* and carefully guided into place. We waited with anticipation as the Mir crew secured us down with cables and finally popped the hatch. It opened with a slight hissing sound. With that, we squeezed our tired bodies through the hatch and into a festive atmosphere of sorts, where a glass of champagne was thrust into my hand once my feet were on the deck. People applauded as though it was an accomplishment simply to come back alive. My past Navy jobs were never like this. *This* was the way to dive! I took a couple of sips and quickly made my way to my cabin, where I took a well-deserved leak after what had been a sixteen-hour dive. It was then that I discovered a massive bruise all along my right side, no doubt the result of my squirming to see out the Mir's viewport.

Mir II was soon recovered and we all foraged though their recovery basket while inspecting the few artifacts they had grabbed from the wreck. Our treasure so far consisted of a few darkened silver coins, dated from the late 1700s to the early 1800s, and a couple of handblown bottles. But there were apparently many more artifacts to recover, such as the remainder of the coins, stacks of plates, and who knows what else. While it certainly didn't sound like a Spanish ship, I felt good about several things. First, Mir II had found the wreck site only 250 meters from where I said it would be, which is pretty good, considering all of the inaccuracies involved in figuring out the location of a sonar target in such deep water. This is because when we were searching for *Liberty Bell 7*, we had no way to determine the exact location of the towed sonar relative to our surface support ship. In such underwater missions, the side-scan sonar is towed behind the ship on a long, steel-armored tow cable, and in deep water that can be quite a distance. During these operations, the location of the side-scan sonar is calculated using the length of tow cable deployed behind the ship, the ship's speed, the corresponding water depth, and the distance of the target to the left or right of the path of the support ship. It's an educated guess, but a guess, nonetheless. Finally, even though we had not found the rich Spanish galleon everyone had hoped for, we *had* found a shipwreck—in fact, the deepest wooden wreck ever discovered. Now we had to get back to the bottom to fully document the site and see what was really down there.

Once more we tossed Mir I into the seas, allowing the grip of water pressure to attack our sub. This time my two fellow submariners were Victor Nischeta, reported to be the best Mir pilot on the *Keldysh*, and Mike McDowell, a likable Australian holding the long-term charter to the research ship. It was Mike's company, Deep Ocean Expeditions, that routinely took well-heeled adventurers to the bottom to explore the likes of the *Titanic* and *Bismarck*. From what I knew, the outgoing McDowell had pretty much done it all, from expeditions to Antarctica, to diving on the Rainbow thermal vents near the Azores, to exploring famous shipwrecks. Victor was a quiet man, not ever saying a whole lot. However, he was obviously a very talented individual: Not only was he a superb submarine pilot, but he also showed a flair for editing the videotapes churned out during the numerous *Keldysh* operations.

Diving with Victor was akin to being in a study hall with a substitute teacher. Not that Anatoly was overbearing at all, but with Victor driving, it was like playtime with the boss on vacation. Unlike the first dive, this time I was prepared. I had extra clothing as well as a bottle of Advil to ward off the possible carbon dioxide migraine. We huddled in the cold.

During the first 1½ miles of our descent, we soothed ourselves with the Celtic music of Loreena McKennitt. Once past the halfway point, though, it was time for action with the hard-edged rock of Vertical Horizon. We nailed the bottom this time very close to the target, and in only twenty minutes, the ghostly shape of our shipwreck loomed out of the darkness.

From bow to stern, what was left of the ship was about 70 to 80 feet long and looked like some sort of massive undersea animal that had been opened up, exposing its ribs. The copper sheathing holding what was left of the timbers surrounded the hull, its surface green with corrosion. Rotting timbers were strewn around the area, many of them anchored to the bottom amid small rivers of rust. What looked like the remnants of a few books gently flapped in the currents, the documents appearing as flowers of waterlogged pulp. A dozen or so ceramic plates remained stacked on the deck near the ship's stern, never to be used. After we documented the location of all the artifacts, Victor parked the Mir right at the stern facing inboard of the wreckage. We then started working on the pile of coins, which looked more like a heap of discarded trash.

Amid the clicking sound of hydraulic valves, Victor whipped out the Mir's starboard manipulator like a Russian gunslinger, collecting hundreds of silver coins with a homemade aluminum scoop. They showered into one of our recovery baskets as he positioned the arm to recover more valuables. It was then that we all saw the gold box.

"What's that there?" I asked, with my face glued to the viewport.

"It almost looks like a gold bar . . . or some kind of box," Mike replied.

Whatever it was tumbled into the basket on top of the growing pile of Spanish coins of the realm. Victor continued working as we were joined by Mir II, making its presence known by churning up a large cloud of mud. After hovering above us like a spacecraft, Mir II crawled up the hill, leaving a trail of swirling sand. (The Mir II crew would eventually find sections of the ship's mast, complete with intact canvas sail cloth.)

It was then that I saw it: a small triangular shape propped up against a soggy wooden timber. "Look at that! Is that a sextant?"

"I think the guys in Mir II mentioned that they thought they saw one during the last dive," Mike replied.

With that, and after finishing up with our now depleted pile of coins, Victor repositioned the submarine and, ever so carefully, extracted the two-hundred-year-old wooden navigational instrument from the ship's silt-covered deck. I wondered whose hands had used it last.

We also heard that Mir II had earlier seen a boot buried beneath a shattered glass demijohn in the same area. Probing with the tip of the manipulator, we managed to dig the black leather object from its grave. A bright red lobster danced around the item as we slowly maneuvered it onto the nearly full basket. While dimensions are hard to estimate underwater, the boot appeared to be of a small size, but of sufficient height to extend about halfway up a man's calf.

It was then that the whole experience became personal. When I saw the boot, I wondered whose ship we had found. This was the boot of some unlucky man who probably died two hundred years ago when his ship sank to the bottom of the Atlantic Ocean. Did any of his crew make it off the ship alive? Did he possibly? Or did everyone perish in the dark seas, waiting for their graves to be somehow discovered? Was there a descendant living somewhere who would appreciate knowing how their ancestor died?

By now our small cargo hold was getting full; we easily had several hundred coins, the boot, the sextant, and numerous ceramic plates and glass artifacts. We dug through the top of a couple more wooden boxes but could only glimpse some unidentifiable material that looked as though it had decomposed into carbonized muck. Our batteries were running low, and it was time to head back to our home on the surface, 3 miles straight up.

During our boring trip to the surface, I still found it hard to comprehend the age of what we had just seen. When the ship sank, there had yet to be a Civil War, and the United States had existed as an independent country for less than fifty years. For a comparison, it would be like investigating a ship in the year 2192 that had been sunk in 2001. How different would the world be in almost two hundred years?

In the *Keldysh*'s sonar laboratory, now functioning as an archaeological processing facility, we began to examine our loot. Jim Sinclair carefully pried open the lid of the ornate gold box, all of us anticipating what we might find inside. While it was small enough to fit into the palm of your hand, it was very heavy. When I looked inside I saw . . . newsprint?

"What's that?" I asked.

Jim dug out a small magnifying glass. "Newspaper."

A newspaper that had survived two hundred years underwater? Impossible! Yet there it was. However, it was only the day before that we had also been shocked to recover an intact sheet of silk fabric, stenciled with the words "Not to be sold." This was not a small fragment, but a silk fabric sample about a meter square. I could now make out some lettering on the paper: "Spanish-Town, February 24, 1809," with what looked like a classified advertisement that read "Adam, a Mundingo, 5 feet 9½ inches tall, right shoulder, has an ulcer on his left side, Mrs. Johnson, Kingston, committed by William Barnes, Esq." A sales ad for a slave, no doubt.

Jim used his dental pick to carefully open the newspaper folds, all the while keeping everything under the water and out of the air. Carefully stored inside the folded-up newspaper were thirteen glimmering gold coins. *Gold!* The captain's private stash, perhaps?

"Looks like we've got some Portuguese escudos," Jim remarked. We stacked the gold coins, all of them dated in the late eighteenth century and never to be spent by their owner.

It was then that my attention was drawn back to the black leather boot. I pulled it out of the small tank of water and removed my own shoe, comparing the lengths. It looked like about a size 9. The captain was obviously not a big man.

Having accomplished our objective, we steamed around the area during the rest of the last night, recovering several navigational beacons that had been earlier deployed on the bottom. With that done, we headed north and back to Bermuda, with all of us trying to figure out how to get our recovered items back to the United States. How do you tell customs that the silver and gold coins you're taking home just came from an unidentified and abandoned sailing ship sunk in 3 miles of water? That's what lawyers are for.

In the end, we recovered over thirteen hundred Spanish silver coins, about half of them identified as eight reales, or later versions of the well-known pieces of eight. We also collected two flintlock pistols (one complete with a brass barrel), a fragile hourglass, two wooden navigational instruments, numerous handblown glass bottles, a ballast stone, a handle for a wood saw, a fragment of a coin box with the name "ROXBURY" stenciled on it, and many other well-preserved artifacts. Yet even if we sold the lot at auction, it would never cover the cost of the expedition.

To this day the ship remains unidentified, and that haunts me. Was the vessel simply taken by a storm? Was there a small lifeboat that the captain and crew managed to use to save themselves? Or did everyone drown as the ship went down? Even though we did not find the rich Spanish galleon everyone was hoping for, we did locate and partially salvage a two-hundred-year-old wooden shipwreck in waters 3 miles deep; something never done before.

Another thing we did was to prove once again that marine archaeology can successfully coexist with commercial salvage operations. The academics who portray commercial salvors as amateurs who destroy underwater archaeological sites weren't on the *Keldysh* or in the two Mirs to see firsthand how careful we were to document the location and orientation of all artifacts before disturbing the site. In addition, all our recovery work was guided by standard archaeological practices as much as possible, taking into account the nature of the deepwater environment we were working in and the technology involved. As Jim Sinclair had put it so clearly during an earlier briefing, this was "cutting-edge" underwater work.

Rather than being sold off at the first opportunity, all of the artifacts recovered from the Atlantic Target were stabilized and cleaned for future study at the South Florida Museum of Natural History—something that would have been impossible without the support of commercial sponsors. I subsequently learned that several years later they were sold to a private collector. One of the gold coins alone went for about $85,000.

The job was not profitable for me, and my legal agreement with Atlantic Sands LLC was ignored. All I received out of the deal was the experience of diving with Anatoly and a single silver eight real coin. So much for treasure hunting.

After diving in the Mirs, I went back to Maryland and my dissolving marriage. I was also offered some short-term employment by Phoenix International, the company that earlier supplied the Remora ROV for our ill-fated 2000 USS *Indianapolis* search. Their commercial division had purchased a Nereus cable burial vehicle from a Spanish company and needed help to get it working. My participation was supposed to be part-time and last about two weeks. I ended up working for them for over twenty years.

13

2010: Air France Flight 447 and *Bluetail 601*

uring the time following the Mir dives on the Atlantic Target shipwreck and the start of the search for Air France Flight 447, I participated in twenty undersea operations ranging in depth from a paltry
200 feet to about 17,500 feet of seawater (the deepest one a classified target object in the Atlantic Ocean) with the CURV III remotely operated vehicle (ROV). This was a span of nine years. The time spent at sea ranged from
a week or so to almost two months, all on either a commercial or US Navy
salvage ship.

The memories I have of these missions are a mix of impressions, such as
the sight of a caramelized human shin protruding from a flight boot after the
collision of two T-39 trainers that collided over the Gulf of Mexico (there was
considerable fire involved); shifted and mangled tectonic plates off the coast
of Indonesia following a tsunami; recovering the crab-infested remains of an
Australian Special Air Service Regiment trooper from the Black Hawk 221
helicopter crash while his best friend watched over my shoulder; sighting dozens of unexploded depth charges and torpedo warheads off Hawaii during
a search for a missing US Coast Guard SH-65 helicopter; and the image of a

crashed USCG C-30 aircraft lying on the bottom upside down with the entire left wing folded over on top of a crushed cockpit following a collision with a Marine Cobra helicopter.

Somewhere in there we did a survey of the aircraft carrier USS *America* (a Kitty Hawk–class CV-66), which was scuttled in 2005 during a weapons test. The entire bridge and primary flight control area had been neatly cleaved off the flight deck, and we almost lost CURV III after it somehow got stuck inside of the hangar deck (fortunately not on my shift; we got the vehicle unstuck). I remember plopping the CURV vehicle right on top of one of the catapults.

When Air France Flight 447 went down in the Atlantic during a flight from Brazil to France, I was working away on various Navy operations with the Deep Drone ROV near San Diego. We had done a lot of work near Coronado, and I enjoyed the place. Everywhere you looked there were signs warning you of one cancer risk or another; they would have done better just to check the ice machines at Burger King for bacteria because we got sick every day from the iced beverages.

It was actually a pretty good gig. We'd camp in the Best Western in Coronado and pile into one aging Dodge Caravan for the eight of us. Mike Unzicker, the project manager on the job, was a notorious tightwad who somehow assumed that he made the company look better by saving the Navy money. The truth was that the Navy paid the same amount, whether we stayed at the Best Western or the Hotel del Coronado.

The best part was that I could walk to the Little Bar down the street, where one of the bartenders entertained us with magic tricks. I could also hit the Albertsons market for food and save money on per diem.

During that evolution, as the SEALs call them, we were working not far offshore recovering an ALFS (airborne low frequency sonar) dipping sonar, surveying an old submarine target, inspecting acoustic arrays, and cleaning off the submerged docking target for the submarine rescue guys. It was fun.

The Air France jet had already gone down right about the time we mobilized for the San Diego job, and we were all wondering how long it would take for some of us to be called away to look for the thing. The problem was, as is usually the case, the airlines and governments who control such disaster efforts always underestimate the technical challenges related to finding a crashed airliner in over 4,000 meters of water. They always assume their own

Navy could handle it. And by the time they face reality, it could be two weeks after the crash. Unfortunately, during their entire period of indecision, the acoustic beacon would be pinging away, using up the valuable battery power until it died thirty days after being immersed in seawater.

We knew little about what transpired on the flight, other than it had gone missing during some bad thunderstorms. If the plane had not shown up anywhere, one thing was certain: everyone on board was dead and must have experienced sheer terror for the last few minutes of their lives.

Eventually we got the call, and a small group of us, along with our soggy luggage, were spirited away from the ARS combat salvage ship (yay!) and shipped to shore to catch an evening flight out of San Diego. We and another group of Phoenix International operations personnel were to meet up in Miami for the long trip down south. We flew down to Rio de Janeiro, got well lubricated during a layover, and in short order were headed to Natal, Brazil. By the time we got to the resort hotel it was 2:00 a.m., and then the real drinking began—for me anyway. It helped knock me out for some well-deserved rest, and by 11:00 a.m. we were on our way to the port to offload our trucks and start installing our gear on the *Fairmount Glacier*, an ocean-going tugboat. It was a frenzied mobilization as usual, with time being the critical element. We got most of the work done in darkness, managed a few more hours' sleep back at the hotel, then were quickly on our way along the plane's flight path.

There were two towed pinger locator systems deployed for this search: ours (the TPL-40) and the PLS-20 on the *Fairmount Expedition*. The bad part was that the French team on the *Expedition* had all the bathymetric data, a decision that would come back to haunt us in the coming weeks. Worse yet, the Bureau d'Enquêtes et d'Analyses (France's version of the National Transportation Safety Board) was running the search operation, despite their lack of experience with such things. At least the NTSB usually listened to our advice. The BEA had their own way of doing things.

My introduction to pinger locator systems was being told to make the TPL-40 work. I admit to not liking the system. I felt it was too complicated and employed an analog-to-digital signal conversion that degraded the fidelity of the signal. It was even worse when I opened the oil-filled electronics compartment, which looked more like a high school science project than a towfish that cost several hundred thousand dollars to develop. I did what I could with it by modifying the pressure compensation system (used to balance the interior pressure of the electronics housing with the ambient environment) and changing the way in which we detected pingers to begin with.

Pinger locator systems are essentially underwater microphones (i.e., hydrophones) and electronics used to detect the presence of a very select band of ultrasonic frequencies, in our case ranging from 3.5 kHz up to 45 kHz. Most flight data recorders have acoustic beacons on them that transmit a 37.5 kHz pulse once per second (1 Hz) and have a life of about thirty days once activated. The vehicles are towed much like a side-scan sonar on the end of a long cable behind and below a surface ship. When I started working with the TPL-40, the surface operators detected pingers by simply listening for the audible signal, or sometimes monitoring a small analog gauge that would indicate when a signal with the correct frequency was heard. I thought this was a primitive way to do the job, so I added a small laptop computer to the system that ran a spectrum analyzer application that added a visible component to the detection technique, one where it would be easier to see the repetitive nature of a pinger pulse. It also allowed us to record the signals using WAV files and eventually became the standard search method for pinger locators, at least those operated by the Navy. In addition, using the spectrum analyzer, we were able to accurately measure the strength of the signal and better log the location of the peak signal strength.

But simply knowing where the pinger was loudest on a search path only isolated the signal in one direction. To determine the geographic location of the beacon required perpendicular lines; if you did enough of these tracks you could eventually isolate the location of the pinger to within 150 to 200 meters—plenty good enough for an ROV with a scanning sonar to find.

In a period of hours, we had all our gear secured to the deck of the *Fairmont Glacier* and were headed to the crash site. There were two shifts made up of three people each. The makeup of the personnel was "not optimum," as they say in operational parlance, because first, we only had three people instead of the normal four. In addition, Phoenix had sent Donna, who had zero field experience. But she made up for it with her intelligence and willingness to follow orders. After some instruction, I decided to have Donna operate the blue deck winch, which held about 7,000 meters of double-armored tow cable. That way, Frank and I could handle the most difficult part of launch and recovery, dragging the 75-pound towfish across the deck and getting it in and out of the water. As usual, I was stuck running the night shift. The worst part was that I was the most physically fit of the shift, so I had to deploy and recover the search vehicle by myself on the fantail. The stress that single task put on my lower back would doom me to constant back pain for the rest of my life.

I can't remember exactly when we arrived on location, but it took several days to get to the accident site, roughly in the middle of the North Atlantic on the flight path from Natal. By June 16, we were searching on line K-1 and soon had the towfish deep underwater with 3,000 meters of cable out while nursing an intermittent data link, unfortunately normal fare for the TPL-40 with its optical fiber connections. To make it worse, the problem was pressure related, which made it difficult to troubleshoot.

Maneuvering a search vehicle is not an easy exercise in ship handling because there are so many ways for it to go wrong. First, the ship needs to steer a generally straight line while dealing with the surface winds and current. Sometimes this means that the bow of the ship is pointed many degrees from its actual path as it crabs into the wind. In addition, once each search line is completed, the ship must execute a long and boring Williamson turn, which can take several hours while we monitor the depth of the towfish and keep the overboarding sheave from banging against the inside of the A-frame. In deep water, just making a 180-degree turn can take as long as twelve hours, depending upon who is directing the ship. Some ship drivers are better than others, and when you get a good helmsman, you want to hang onto them (they're usually better at ship handling than the captain).

By June 27, we were well into the search and so far had heard nothing that sounded like a pinger. Also, we didn't understand why we were searching where we were. The BEA had us far to the north, where the vertical stabilizer had been found after more than a week of drifting in the ocean currents. The plane would not be there.

What passed for bathymetric data on our ship, as opposed to the *Fairmont Expedition*, was a hand-drawn square on the wall with various circles drawn inside of it, each of which contained numbers representing the water depth in particular areas. While the chart represented an area 20 by 40 square nautical miles, there were no geographic references to where any particular bottom depth was located, and the ship's bottom sounder didn't help us. We had just completed paying out cable and were on line 34E when it happened.

June 28, 2009, 0200 hours ship time. I could barely keep my eyes open as the shushing sound of the towfish's background signal was true white noise, enough to put anyone to sleep in the dark hours of the night shift.

Every now and then, Frank would prod me with his elbow. "Hey, wake up, you're falling asleep again!" When that happened, I would sit erect for a while, then eventually slump back down in the most worn-out and uncomfortable office chair known to humankind. You would have thought that it alone would keep me awake. But it didn't.

The most exciting thing to happen on the shift was being cut off in an ocean-going traffic jam by the *Expedition*. Of course they made us give way. The only signal anyone heard was from the *Expedition*, and that turned out to be a screwdriver rolling around in a drawer.

We had just hit start of line with my eyelids partially open when the loudest sound I had ever heard come out of the Bose speakers erupted and filled the interior of the van—the sound no one ever wants to hear—that of the search vehicle striking and being dragged across the seafloor.

"We've hit bottom!" Frank shouted, as we both leaped from our chairs and headed to the deck. It was bad. The winch was rendering with a loud screeching sound, which meant that the towfish had stopped forward motion and was being held against the bottom while the ship steamed ahead, stripping cable from the massive reel.

We quickly started paying out cable in an attempt to save everything as I called the bridge and told them what was happening. To make matters worse, the winch's electrics overloaded and shut down.

"RESET!!! RESET NOW!!!" Frank screamed. Finally, whatever giant claw was grasping our towfish finally released it and we were able to haul in tow cable. I went back into the van and saw, very much to my surprise, that we were still receiving data from the TPL-40. I couldn't believe it. Maybe it was not so bad after all . . .

But it was. What I saw hauled out of the water was nothing but a mangled piece of stainless steel tubing now shaped into an inverted V, with shattered hydrophones, missing fins, and sliced cables. The job was over.

Five minutes later, I grabbed Mike's shoulder and shook him awake. "Mike! Job is over, dude. The towfish is toast. We hit bottom." A groan was all I got for a response as he slipped on some work clothes and we headed to the deck.

I must give us a lot of credit for saving the day with a gas blowtorch, a sledgehammer, and some wooden dunnage. It took hours, but after heating up the center of the burned and crumpled steel chassis, we managed to beat it into shape, or close enough. This repair had to rank up there with my in-line splice of a TROV ROV's umbilical in the Gulf of Mexico in 1978; it was awesome. And the thing still worked, though the directional hydrophones were smashed. We didn't need them anyway.

We did the rest of the job with our slightly bent towfish, prompting a BEA engineer to draw a hilarious cartoon showing our mangled search vehicle leaping out of the water like a harpooned swordfish. At least the BEA frogs on the *Fairmont Expedition* thought it was funny. We were a bit embarrassed.

By the end of the job, we had heard nothing from the pingers that went down with Air France Flight 447. That was because, unbeknownst to us, neither of them was working. As was discovered during the later salvage, one pinger was buried in the mud and the other one had been flooded. To make matters worse, on their last search line, the tow cable dragging the PLS-20 behind the *Fairmont Expedition* inexplicably failed and the vehicle was lost at sea. That is just what happens sometimes with these operations. It goes with the territory. Sometimes you get them back; sometimes you don't.

As a result, we never found Air France 447, at least not on that trip. It ultimately took three autonomous underwater vehicles and a towed sonar to get the job done, after two years of searching. Unfortunately, the BEA was like most investigative organizations—out of their element. They knew a lot about crash investigations but not how to search in deep water.

By the time we got back from Brazil, our ability to locate black boxes on the bottom of the ocean had been decimated. The TPL-40 looked as though it had been run over by a tank, and the PLS-20 was on the bottom of the Atlantic. Clearly, we needed a stopgap measure until we could get another search vehicle. Consequently, I took it upon myself to cobble together one we could use in the short term.

We still had the topside receivers from the lost towfish (the PLS-20). While they were primitive analog systems dating from the 1980s, they did work. All I needed to do was create the underwater portion of the system. I rummaged through the shop and found enough spare circuit cards that could power a hydrophone and filter the audio signal. I even managed to find an old

Massa hydrophone, which probably cost a fortune when it was first purchased. Unlike newer designs, it was oil-filled and had a flexible rubber boot. But after I took it apart and made some adjustments, it proved to be very sensitive. I also needed a pressure housing to protect the electronics from the crushing pressures of the deep ocean.

At that time, the CURV III ROV was being decommissioned and readied for display at the US Navy Museum at the Washington Navy Yard. Fortunately, it contained a vintage aluminum Osprey low-light camera housing rated for 6,000 meters just waiting to be appropriated for my little project. I opened the electronics bottle, stripped out the electronics, and had a local machine shop thread a penetrator into one side to hold the hydrophone. Following a couple of weeks of fabrication and for a pittance, I ended up with a usable underwater housing that held the electronics and hydrophone. Best of all, it worked, very well in fact.

The final task was to build some sort of towfish or frame to hold the electronics bottle so it could be towed underwater. Again, for a minor contribution of Navy funds (technically, to buy spare parts), I designed and built a tow body made from a combination of high-strength plastic, thin-wall aluminum sheeting, and some heavy gauge stainless steel because I needed the thing to be strong. In the end, for a few thousand dollars and some of my labor, we had a new towed pinger locator system, one that would later be used to find a missing Northrop Grumman airborne early warning (AEW) E-2C Hawkeye, *Bluetail 601* (Bureau No. 165508) of the VAW-121 squadron. This aircraft was destined to crash into the Arabian Sea following a mission into Afghanistan. Of course, at that time I didn't know any of that. No one did. But this new contraption, named TPL-21, eventually intertwined two lives: mine and that of a young Ukrainian immigrant who wanted to become an American and learn to fly. And fly he did.

Lt. Miroslav "Steven" Zilberman was born on February 20, 1979, in Kiev, Ukraine, not far from the charred toxic remains of the Chernobyl nuclear reactor. His nickname was "Abrek," the name of one of two primates that were sent into orbit by the Soviet Union on the Kosmos 1514 flight. Fearful of the radiation leaking from the Chernobyl reactor, his parents, Boris Zilberman and Anna Sokolov, decided to emigrate to the United States in 1991 to make a better life for themselves. Ironically, they were also concerned that their bright and energetic son might be forced into the Soviet Army, only to see him later join the US Navy. Mother Anna was a biochemist and worked for the Kiev Institute of Hygiene after earning a master's degree and PhD in the field. Boris was an accomplished construction engineer. Anna Sokolov was later "shocked" when she learned that her only son had decided to join the US Navy. "We were afraid of the military service because it was awful for Jewish people in the Soviet Union."

Young Abrek adapted to life in Columbus, Ohio, and was by all accounts "incredibly smart." He picked up English quickly, and his English teacher Marilyn Rofsky recalled that Abrek once interrupted her class by asking for fifty cents to buy a soda. When she confronted him and asked him what he was doing, the twelve-year-old evenly replied, "Well, I am thirsty." Over time, Rofsky helped Abrek adjust to life in the United States and along the way discovered that he was strong-willed, knew what he wanted out of life, and was willing to do what he needed to get it. He also soon grew into a capable, driven, and handsome young man.

The republic of the United States is an experiment made from the toil of immigrants who, over time, created an American culture, a mishmash of ethnicity and European history. Boris, Anna, and Abrek embraced our country and, over time, became Americans. These American values and beliefs, coupled with his Ukrainian background, drove young Abrek toward an ethic of hard work and a sense of purpose.

Once Abrek made it to high school, he met Katrina Yurchak, a Torah Academy student who later became his wife. Although he was accepted to Ohio State University, he decided to join the Navy instead and attend Rensselaer Polytechnic Institute in Troy, New York. Possibly, part of his motivation for doing so was that he was inspired by his maternal grandfather, Grigory Sokolov, a Russian fighter pilot during WWII who was shot down twice and held for eleven months as a prisoner of war.

Abrek Zilberman didn't want to rely on his parents and paid his own way through college; the US military made that possible. Even with time dedicated to the Navy, he still managed to earn a bachelor's degree in computer science from Rensselaer in only three years. Because his ultimate goal was to become an emergency room doctor, he spent much of his time studying books on organic chemistry. Incredibly, Zilberman turned down an appointment to the US Naval Academy so he could marry his girlfriend, Katrina. Following graduation from college, he was commissioned in 2003 and eventually ended up in Carrier Airborne Early Warning Squadron 121 (VAW-121), the Bluetails, as-signed to the carrier *Dwight D. Eisenhower* (CVN-69). It was not long before he was an accomplished pilot flying the Northrop Grumman E-2C Hawkeye, an all-weather tactical airborne early warning aircraft, originally produced in 1960 with the nickname "Super Fudd."

The E-2 Hawkeye aircraft Zilberman flew was developed in the early 1960s by the Grumman Aircraft Corp., a company with a well-earned reputation for designing and building strong airplanes. Zilberman's E-2C version incor-porated the latest advances in radar and avionics, and the overall purpose of the aircraft was to collect and distribute tactical battlefield information back to Navy ships. After several decades of use, the C version was upgraded to incorporate a new eight-bladed propeller system (NP2000), which resulted in better fuel efficiency, less vibration, and easier maintenance because the in-dividual blades could be removed for maintenance or repair. Later upgrades included a glass cockpit and software updates.

Zilberman's Hawkeye was a large airplane with an empty weight of 40,200 pounds and two massive Allison T56-A-427 engines, each developing about 5,100 horsepower. The plane was over 60 feet long with a wingspan of 80 feet. The distinguishing radar dome was 24 feet in diameter and rotated at about 4 rpm; the aircraft was in many ways a smaller version of the Boeing E-3 Sentry AWACS (airborne warning and control system). It was an exceptional airplane with a good reliability record with VAW-121, but . . .

Every mechanism devised by humankind is inherently flawed in some way. Why? Because human beings are not perfect, and everything created by them has potential flaws. Over time, these imperfections, which often lead to di-sasters, came to be known as gremlins. These are the same ones referred to time after time by pilots worldwide. One Royal Air Force fighter pilot, Roald Dahl, flying with the 80 Squadron, wrote of the existence of these phantom creatures in his first book, *The Gremlins*, following his experiences flying for the RAF during the war:

> It was some time during the Battle of Britain, when Hurricanes and Spitfires
> were up from dawn to dark and the noise of battle was heard all day in the
> sky. . . . It was then that the first gremlins were seen by the Royal Air Force.

These are the mythical creatures that, without permission, fly along with and inhabit every airplane ever flown in history. No one knows what they look like or why they're there, but they are real, there is no doubt.

While Roald Dahl made gremlins internationally known, many other returning pilots have attested that they saw creatures tinkering with their equipment. One crewman swore he saw one before an engine malfunction that caused his B-25 Mitchell bomber to rapidly lose altitude, forcing the aircraft to return to base. How can I be so certain they exist? Because I made a living for nearly fifty years cleaning up the results of their work.

Strangely, the results of a gremlin's work can make amazing and great things happen as well, such as my finding Gus Grissom's Mercury spacecraft on the first dive in 1999, or managing to detect the pinger of a C-2A Greyhound aircraft lost in 5,500 meters of water on the first line in 2018, using a damaged tow cable. But more often than not, these little buggers create nasty situations, such as the eroded O-ring seal on the space shuttle *Challenger*; falling ice from the external fuel tank on the *Endeavor*; bad wiring in the scavenger pump of a center fuel tank in TWA 800; a fused level sensor in a liquid oxygen tank in Apollo 13; and defective electrical wiring in the Apollo I command module.

On the morning of March 31, 2010, at 9:20 a.m. in the Arabian Sea, there was a gremlin curled up and lurking inside the right Allison T56-A-427 engine of Zilberman's E-2C Hawkeye as he took off in support of Operation Enduring Freedom with the call sign Bankroll 61. Unlike the imaginary creature depicted in the 1963 *Twilight Zone* episode "Nightmare at 20,000 Feet," this one was real and damn well had his number. Zilberman would never become the doctor he had dreamed of becoming, but he did gain mythological status among his fellow pilots for the decisions he made and the actions he took that day.

What follows are brief excerpts (in italics) from the May 25, 2010, investigation report, plus commentary.

From: Commander, Carrier Strike Group EIGHT, 25 May 2010: COMMAND INVESTIGATION INTO THE FACTS AND CIRCUMSTANCES SURROUNDING THE VAW-121 AVIATION MISHAP AND FATALITY THAT OCCURRED ON 31 MARCH 2010 IN THE GULF OF OMAN.

On board *Bluetail 601* that day were Lieutenants Jeremy Arnott (copilot and Zilberman's roommate on the carrier), Edmund Poynton, and Richard Holt. Arnott later remembered Zilberman as an "outgoing wisecracker" who would sometimes send out prank messages to other officers when they forgot to log off their computer. "At first glance at Steve, it might be difficult to see him focusing on anything when he always had so much energy and would bounce from one topic to the next. . . . I was amazed that in private he brought all of that energy and brought it under control and focused it on ambition," Arnott recalled.

All four of the men were eager to fly.

Weather at launch, Mission and Recovery was SKY: FEW-SCT 250-300 / VIS UNRST, 5-7 DU HZ SLANT, SFC 060, WINDS (KT) 20-25 G30 . . . Man-up and preflight were uneventful . . .

Zilberman was all business when it came to flying because he knew, like any aviator, that his very life depended on it. With the Hawkeye attached to the steam catapult's shuttle and all the crew members strapped in place, the massive *Eisenhower* turned and sliced into the wind, while Zilberman pushed the throttle levers forward to almost full power . . . his fingers tense on the controls, his eyes glancing at the instruments, watching the flight deck rise and fall, until the shooter judged it just right. The captured 60-foot-long aircraft howled with power and vibrated and shook like a restrained animal, waiting for the gate to fall.

As the four men felt a massive shove at their backs and heard a loud whooshing sound, *Bluetail 601* exploded down the smoke-filled deck like a ballistic missile, until it hit the end of the bow and, with propellers clawing at the air, entered the realm of free flight. The dull-gray-colored vehicle with "601" on the tail slowly disappeared into the distance and headed for Afghanistan.

No information has been released describing what was or was not accomplished during their flight, except that it was "routine." That, unfortunately, was about to change.

The initial RTB check-in with Strike was conducted approximately 50 nm from ship at an altitude of 24K. . . . It was passing through 17K when a "R ENG OIL LOW" caution appeared on the instrument panel in the cockpit.

Seeing the glowing indication on the aircraft's warning panel, Arnott tersely asked if anyone else on board could see anything wrong with the right engine. Everything had been normal up to now, until they descended through an altitude of 17,000 feet only 50 miles from the *Eisenhower*.

The co-pilot inquired of any visible leaks from CICO [the combat information center officer,] to which CICO responds in the affirmative and voices his concerns that fluid loss was "dramatic." . . . Oil [was] streaming down the left side of the starboard motor, which they described as "light and slightly orange in color."

Poynton, who was the combat information center officer and on the right side of the aircraft, loosened his straps a bit and peered outside. What he saw was not good. "Affirmative, I can see a lot of oil streaming down the number 2 engine. . . . It looks orange . . ." At that moment, the entire crew knew they were in trouble. How long would the engine last? Would they make it back to the carrier? It was impossible to say, and all everyone could do was remember their training and do their jobs.

Approximately 12 miles from the ship and at 3,000 to 4,000 feet, the pilot advised, "We need to secure the engine."

Zilberman made that decision because of what he was seeing in the oil pressure warning indicators. The oil pressure warning light was strobing, in that two of the warning LEDs, the amber and green ones, were lighting up in an alternating fashion, meaning that the engine's oil pressure was dropping. Things were rapidly going from bad to worse. What the pilot did next was by the book. Zilberman pulled the starboard power lever back to idle and shifted the rudder trim to 20 degrees left to compensate for the loss of power in the right engine. But he didn't turn the engine off—not just yet—because he needed other things to happen before making that decision.

In any twin-engine aircraft that loses power on the right engine, the aircraft will try to yaw to the right, toward the problem. It can sometimes be a violent experience. For example, when an SR-71 Blackbird lost an engine, the pilots instinctively knew which one it was by which side of the canopy was hit by a helmet.

As soon as that power level went to idle, the airplane quickly careened to the right as Zilberman fought to keep it under control by pressing hard on the left rudder pedal. He probably even put in some left aileron to help. Soon, they were only 12 miles from the *Eisenhower*, almost within sight. Crew members came out on deck when they learned that one of their planes was in trouble. The next thing Zilberman did was to put some flaps out to reduce the Hawkeye's stall speed—standard procedure.

> *With the RPM stabilized between 71–73 percent . . . the co-pilot called out, "The engine did not feather."*

All twin-engine aircraft have "engine out" procedures that detail the necessary steps to be taken to safely fly an airplane with only one engine. They vary in complexity, but in the case of engines with variable pitch propellers, one thing is consistent: the propeller on the failing engine must be feathered. In normal flight mode, the props of an aircraft bite into the air at different angles depending upon how much power is needed. But if you lose an engine, the props of that engine are "feathered" so that the edges of the blades do not add drag to that side of the airplane. In the case of *Bluetail 601*, for whatever reason, the propeller on the right engine refused to feather, adding so much more drag to that side of the aircraft to make it literally uncontrollable in flight. With the right engine out, the propeller blades unfeathered, and with the radar dome on top, Zilberman was not flying an airplane anymore—he was flying a brick.

> *All five methods to electrically and mechanically feather the engine failed, this includes 1) Electrical T-Handle Feathering; 2) Electrical Condition Lever Feathering; 3) Electrical Auto Feathering; 4) Electrical Feather Override; and 5) Mechanical Feathering.*

Zilberman had to lock his left knee to keep the rudder depressed, try to stay level, communicate with his crew, and make split-second decisions on what to do. They had tried every procedure in the book, all five of them, and they had all failed. He ordered Arnott to take control of the aircraft so he could stand up and jettison the escape hatches; there were two of them, each over a pilot.

The pilot relinquished control of 601 to co-pilot in order to emergency jettison crew door. The co-pilot was required to lock left knee for full rudder deflection and was struggling to maintain wings level.

There was no more time for screwing around. If anyone was to make it back alive, it would not be in *Bluetail 601*. Either Arnott or Zilberman radioed back to the ship that they were bailing out. But they couldn't just get up and leave. Someone had to keep the aircraft under control. Someone's leg had to hold the rudder pedal and hands had to grasp the control wheel; there was no other way. If anyone took their hands off the controls, the Hawkeye would roll over, stall, and crash, killing everyone aboard. Zilberman was the aircraft commander, so as soon as he was done opening their only means of escape, he took over from Arnott. "Pilot to crew, bail out!" They would be the first ones to bail out of a Hawkeye in twenty years.

Witnesses on the flight deck visually acquired aircraft immediately following the "601 is bailing out" transmission on the tower frequency.

While they were within sight of the flight deck, no one remembered seeing the crew exit the aircraft. As Poynton later wrote in *Approach* magazine, "He sat on the edge [of the door,] his feet hanging in the air. . . . Holt slipped on the edge of the hatch and fell out headfirst."

They were falling quickly down to 1,000 feet above the water, the minimum safe bailout altitude, as Arnott stood up to get out of the airplane, hoping his friend would quickly follow. As Arnott remembers, "Steve looked over and told me firmly to 'GO, GO, GO!' . . . He wanted to make sure I had time to get out safely, and he knew it was his duty to keep the aircraft flying long enough for us to get out safely."

But they didn't all get out safely. Steve Zilberman didn't want to die in those last seconds, and probably spent his last moments trying to ditch the aircraft and get out before it sank. But setting an aircraft down in the water is not an easy task, especially one that has decided to stop flying. Zilberman was now at 11/10ths of control.

Every aircraft has a flight envelope within which the plane can be controlled. It is possible to fly on the edge of that envelope, but all the forces keeping the aircraft under control are balanced on the edge of a knife, and it doesn't take much to push it off. It is much like famed Finnish racer Kimi Räikkönen driving at 10/10ths. You are literally on the edge of controllability and using every fiber of your being to keep everything in balance. You feel the physical feedback from the rudder pedals, the control wheel, the G forces on your body, and what you're seeing with your eyes with respect to spatial orientation. It is driving by the seat of your pants. Likewise, the pilot feels it through their body and even uses their hearing to absorb every bit of sensory information they can. But this is almost impossible to sustain because a pilot's body eventually becomes fatigued, and once that happens, all it takes is a small error in control input to push the aircraft over the edge. Zilberman did the best he could. But eventually, after giving his flight crew enough time to bail out and save their lives, he fell off the edge of the knife and departed from controlled flight. It was a valiant effort, but try as he might, the Hawkeye simply got away from him; and once that happened, there was no going back. He crashed.

> *Witnesses state that passing the Island there is a slight drop of the left wing and aircraft returned to wings level with a noticeable increase in rate of descent. There was another dropping of the left wing slightly and the aircraft departs controlled flight and impacts the water. . . . Witnesses did not see aircrew bailout.*

The impact of an aircraft at any speed is a violent event, one where the human body, even though it is an amazing creation, does not fare well. In a high-speed crash, any person is terribly fragile.

The Naval Air Training and Operating Procedures Standardization program procedure for "fail to feather" says for the pilot to maintain at least 135 KIAS (knots indicated airspeed). That airspeed equates to Zilberman's Hawkeye covering 228 feet during every second of flight. That is almost four aircraft lengths per second.

Seawater is an incompressible fluid, and hitting it at 135 knots is more akin to running headfirst into a wall of concrete. Zilberman's E-2C Hawkeye smashed into the ocean, and in quick succession, several things happened: The airplane impacted in a left wing and nose down orientation. When the nose hit the water, the entire cockpit section—everything in front of the wings—disintegrated and was reduced to a mangled section of shattered pieces of aluminum, all initially held together by miles of electrical wiring and hydraulic lines. Once the wings hit and decelerated, both turboprop engines shot forward like cannonballs because with the wings stopped, the structure holding the engines in place failed. The aircraft's main fuselage and tail assembly were ripped away and were no longer part of the plane.

Loss of Government property TBD with respect to aircraft 601 and salvage operations. Additional property loss, two Radio Sets valued at $6,986.00 each and four Storage Batteries valued at $182.00 each. $14,700 total.

Once the breakup was complete, what was left of the Hawkeye were five major sections of wreckage: the cockpit, fuselage, wings, both engines, and the tail. These quickly sank and started a long decent to the bottom of the Arabian Sea, under a rapidly spreading slick of kerosene JP-4 jet fuel and hydraulic oil, sprinkled with small bits of insulation, possibly a life raft, some flotation vests, and anything else that would float. Zilberman had been swallowed by the abyss of the Arabian Sea. By then, the only living thing left on *Bluetail 601* was the single Dukane acoustic beacon (designed to withstand a 3,400 g deceleration for 6.5 milliseconds), which was sending a 37.5 kHz signal every second into the deep waters, like a beacon marking Zilberman's grave, begging to be found so that nothing like this could ever happen again.

US Navy Lt. Steven "Abrek" Zilberman would not be forgotten, not by his fellow fliers, his family, the Navy, nor by me once I learned his story. Kate Wiltrout of the *Virginian-Pilot* wrote in April of 2010, "Had a lesser pilot been at the controls of Bluetail 601 last Wednesday, there might have been four memorial services this week instead of one." She was so right.

When Zilberman's plane crashed, the word spread throughout the Phoenix shop in Largo, Maryland, as such things usually do. Our managers began cranking out cost estimates and examining shipping expenses, air cargo flights, likely support ships, the equipment required, and what people were needed to support a search. Usually there's an initial inquiry about the fate of the crew, then shoulders shrug, people comment that "things happen." At the time, I had no idea about Zilberman or his story, and it would have been unusual for me to even care because I had other things on my mind, such as

Would I be going? and if so, *Would the Navy let me use the towfish I built, or would I be forced to use the TPL-40?*

In fact, even if an aircraft is fitted with a pinger, it is not necessarily a given that there will be a pinger search. If we feel that the position is pretty good, many times we will jump right to a side-scan sonar search. In the case of *Bluetail 601*, logic prevailed because the water was too deep for the SWISS (shallow water intermediate search system, a towed side-scan sonar) given our available cable length, but shallow enough to drag a pinger locator because they ride much farther off the bottom and also give us the best coverage. The only other sonar that could reach the bottom was Orion, which is a huge and costly system to mobilize, requiring far more people. So, we decided to keep it simple by sending me, Mark Bender, and four other Phoenix personnel, making it three people per shift, just enough warm bodies for such an operation. The only bad part was that we would be stuck using the older gray winch and the cable counter didn't work—or at least we would find out it didn't once we got there. That meant we had to guess the amount of cable we had out, which was not a big deal because we'd be putting it all out anyway to reach 10,000 feet. Our point of mobilization was Manama, Bahrain, and the support ship was the USNS *Catawba*, an aging but well-maintained Military Sealift Command T-ATF fleet tug.

While the rest of the crew felt otherwise, I found the idea of going to Bahrain exciting . . . sort of like heading downrange, or being the tip of the spear. We were going where the action was, the Middle East, with all the mystery and danger we imagined. We would not be unarmed at sea, either. The Navy was supplying a security detail, a small group of reservists who were not exactly excited about the idea because they had thought they were going home to San Diego—not to another ship. They had just pulled duty protecting a tanker and were sick of the place.

The king of Bahrain, Hamad bin Isa Al Khalifa, is a Sunni Muslim. By the time we arrived in his kingdom, he had been the monarch of Bahrain for eleven years, since 1999, initially reigning as the emir, and starting in 2002 as the first king. The country has been ruled by the Al Khalifa dynasty since 1781. What this meant, I would learn, was that if you were a Sunni Muslim, you got all the benefits from living in the kingdom, such as preferential treatment, a good job, extra money, a nice place to live, and so forth. But if you were unfortunate enough to be a Shia Muslim, you were stuck with low-paying jobs and substandard apartments. (Although in fact, most of the service-oriented jobs were not performed by Muslims at all, but Filipinos.)

When we got to the country it was just before the 2011 uprising where the Shias protested the situation. The king of Bahrain has a slightly different way of dealing with protests in his country than we do in the United States. They simply bulldozed the protesters encamped in Pearl Roundabout, leading to numerous arrests and dozens of deaths. It was common knowledge that black-hooded men were raiding the homes of the protest leaders on a regular basis, usually in the middle of the night. These people, most of them college professors and physicians (i.e., educated individuals), simply disappeared, many of them never to be seen again. They were certainly tortured and forced to disclose the identity of the other leaders.

Fortunately for us, when we arrived, it was still safe to explore the city of Manama, which we did. I was surprised to learn that, in addition to the women, there were dress codes for the men. While the women were saddled with hijabs and burkas, the men were discouraged from wearing shorts or even short-sleeved shirts—this in a country with a climate like a blast furnace. I ignored the protocol and observed nothing but a few strange looks; after all, I was a stupid American.

It was oppressively hot in Manama, and there was always a dull yellow pale hanging over the city from dust storms in the desert that flung small particles of sand high into the atmosphere. A massive mosque stood right across the street from our luxurious hotel (all hotels are luxurious in Bahrain) that broadcast the morning prayer like clockwork. There were also small arrows in the top drawer of the nightstand pointing the way toward Mecca. It was a big joke for us to loosen them and point them the wrong way.

The USNS *Catawba* was a good ship, as much as any T-ATF can be such a thing. It was the usual drill: cramped bunks with little or no headroom down in the pit, a head that always stinks, and no place to store your baggage. Worse yet, we had the security guys down there with us and so were left with even less space because all their Pelican (equipment) cases would be crammed in between the racks. In addition, all the mattresses had recently been replaced, and what we would be sleeping on was as hard as concrete. On the other hand, they had won awards for their food, so there was that.

There was an all-female rock band playing at the hotel's club, and it was a relief to be able to smoke there. In fact, you could smoke anywhere and no one cared. My tobacco of choice at the time was Marlboro Lights, and it was important to have enough smokes on the ship because a nicotine fix was critical to dealing with the tension of the job. There's always job tension, and it doesn't matter how well things are going. You're always afraid of failing, damaging equipment, or getting hurt along the way.

While loading our equipment, which had been delayed in customs, I struck up a conversation with a Navy seaman who was assigned to a formidable-looking gunboat on the other side of the pier. I had never seen such a small ship with so many weapons; .50 caliber Brownings, .240 caliber automatic rifles, a couple of Bushmaster 20-millimeter rapid-fire cannons, and probably some shoulder-launched missiles in a deck box. As he took a long drag from his cigarette and wiped the sweat from his forehead, the seaman casually recounted how they had been attacked by pirates on the last trip out to sea. Apparently very stupid pirates, because he allowed as how they all got "smoked." But you never heard that on the stateside news. In fact, there were a lot of things you never heard about back home.

While we waited for our equipment to clear customs, someone among us suggested we go out to see the "Russian band," an apparently hot group of Russian females performing at a nearby bar. So, we headed out into the night to see these Russians, in a large group of course. The place was a hole-in-the-wall a short walk from the hotel, and as we got settled with our drinks in the near-dark lighting, a man in full traditional Arab dress glided in and took a seat at the end of our group as a young waitress followed him carrying a large hookah. It was starting to look like the cantina in the original *Star Wars* movie.

Before they started, the performers greeted all of us individually, with the leader (who appeared to be the oldest) letting her hand linger on mine far longer than necessary. Was this a signal to me of other possibilities with her? Who knows.

So, they started singing, a mix of dated pop songs with a few traditional Middle Eastern songs in the mix to keep the locals happy. One of the women was quite attractive, a tall blond beauty with long legs. She was always out front. None of them played any instruments. They were simply backed up by an aging man playing a music synthesizer.

It was after a few songs that one of their assistants began to walk among the tables offering small silver crowns to us, which, to show our appreciation for their talent, we could take up to the front of the stage and place on their heads. OK. So, what's the big deal? I took one of them and gave it to the tall blond, who bowed her head in acknowledgment and gave me a large smile. As the alcohol flowed, more crowns were placed on the singers' heads, many of them by Ed, a tall, good-looking ex-machinist mate from a fast attack nuclear-powered submarine. Ed was having a blast, running up to the stage like a kid in a candy store and winning admiring glances from the singers.

It was about then that I noticed the lady who was handing out the crowns lurking in the shadows off to the side of the stage. Hmm. Every time someone took a crown up, she made a notation in a small notebook. Then it dawned on me: the crowns were not free. How much did they cost? It was impossible to say, but here we were in a Muslim-majority city, which likely had severe penalties for just about anything, and we were probably running up a huge bar bill of some kind. I leaned forward and tapped Carl on the shoulder. "Tell Eddie I don't think those crowns are free." As Carl whispered something into Eddie's ear, Eddie's large grin slowly faded into an expression of disbelief. I remember seeing him wave the waitress over and say something to her. She said something back. Then I saw his face turn into a scowl as an argument ensued. It was then I discovered that each crown added $60 to your bill, and Ed had put a lot of silver crowns on a lot of Russian heads. Another scam worth noting.

It was about that time that I decided I was feeling tired and wanted to head back to the hotel. I received and paid my bill (the crown tax had yet to be added), quickly got up, and left as invisibly as I could. I think I made it about thirty paces out the door before a small Asian waitress came running after me, shouting, "Your bill! Your bill!"

I turned around and said, "I don't know what you're talking about. I paid my bar bill," then turned on my heel and scurried back to the hotel, hoping that the Russian mafia was not going to be chasing me down.

We sailed from Manama sometime before April 15, according to my log, for a short transit through the Strait of Hormuz and east into the Arabian Sea. Ours being a Navy ship, we always transited through the straits at night to stay out of sight of the Iranians. But it didn't stop smugglers from following us through, using our ship for radar cover.

Most people cannot appreciate the darkness of the open sea. Once you get a few miles from land, your ship becomes an oasis of humanity, totally isolated from the rest of the world. Or at least that's what it feels like. With a darkened ship, as ours was, the millions of pinpricks of light that make up the stars come out, filling the night sky and making you feel insignificant in the universe. You could go out on deck and be in total darkness, and slowly over time, as your eyes adjust, you begin to see a night sky not obscured by light pollution. A dark sky unlike any other, filled with millions of stars and clouds of galaxies.

Setting up a deepwater search such as the one we were about to attempt is a process of putting the ship in the right location relative to the target at the right time relative to deploying the tow cable. Getting 6,000 meters of tow cable into the water takes a long time because it must be done slowly to allow the cable to stabilize behind the ship. Before any of that even happened, we conducted a "set and drift," where the ship stops engines in the center of the area so we can determine what direction the current and wind are moving the ship; then we can compensate for this movement during the search. This we had already completed as we finally started putting the cable out early in the morning of the sixteenth, while steaming slowly in the northerly direction toward what we were told was the crash location. On that morning, it took us about two hours to get 6,000 meters of cable out as we dragged my home-built towfish far below and behind our ship in roughly 10,000 feet of seawater. I remember that we started far south of our set and drift position to give us extra time to get the towfish down to depth.

In the darkness far behind and below the *Catawba*, the towfish was jerked through the water column, listening for a very specific frequency of sound, 35,700 cycles per second. Just that one signal, the one that could lead us to *Bluetail 601*. The pressure on the thing was immense, over 4,400 pounds per square inch. But my modified Osprey camera housing held, even though it had never even been in the water before.

Incredibly, after only an hour of steaming toward Steve Zilberman's crash location, I began to hear something, random and faint. *Tick.* Just a single tick. Nothing repetitive, just a single ping from the deep. Then more. *Tick . . . tick . . . tick . . . tick.* I could see the single pings scrolling down the spectrum analyzer like breadcrumbs, becoming more solid all the time.

Tick . . . tick . . . tick . . . tick . . . A pattern now, at the right speed of one ping per second, and getting louder all the time.

I passed on the news. "Deck, bridge. I am starting to hear something. Pay out about 100 meters of cable."

"Roger that."

Soon, it was coming in solid like a voice from the deep, leading us right to the airplane. *Tick . . . tick . . . tick . . . tick . . . tick . . .* I could see it well now, as a continuous trail of dashes scrolling down the computer screen.

From my log:

17 April

02:45 Receiving occasional pinger hits on spectrum analyzer at 23:46 UTC.

02:51 Audio pinger start.

03:02 Signal detect light w/full gain.

03:31 Pinger detect w/30% gain.

03:46 Pinger faded out, but faintly visible on spectrum analyzer. No signal detect at 100% gain.

That was a good sign, meaning that we were close enough to *Bluetail 601* to hear the pinger and that it was still working as designed. In many cases, we had to complete several passes in an area before we would hear anything at all. But this pinger we heard right off. The rest of the job would be fine-tuning the location.

We turned the ship to the north after doing a long, drawn-out Williamson turn, taking in cable, turning for hours, paying out cable, then getting settled out on a course to the west.

I turned around to face the helmsman. "Slow to 2.5 knots . . . Try to stay on line." He nodded his head in silence. The bright morning sun illuminated the bridge as I strained to see the computer screen.

After about thirty minutes, more signals from the grave. Strong signals, as though Steve Zilberman were leading us to the spot. *TICK . . . TICK . . . TICK . . . TICK . . .* Really loud now, as my log recorded:

17 April

09:02 Pinger audible start.

09:07 Signal detect 100% gain.

09:21 Very strong 25 db+ 06:22 UTC.

We were there now. We had to be close to the airplane. All we needed to do now was make some north–south runs to nail down the crash position.

Another line done. We hauled in cable again, starting another line in the other direction as our aging diesel–electric power unit coughed and protested, straining to pull in about 3 tons of cable. Another hour was gone as we finally got the double-armored cable in to 3,000 meters out. Then a long, boring turn to the north to do a perpendicular line to box the thing in. The *Catawba* careened across the Arabian Sea, dragging over a mile of cable behind, with lookouts searching for pirates, the static from the VHF radio filling the bridge, coffee flowing to keep everyone awake, and the day shift fast asleep down below on their concrete racks without a clue as to what was happening.

Finally, Mark dragged his tired body up from below to get the dope on the situation. "Waz happening?" he asked while taking the first drink of the day of his continuous cup of coffee. Mark drank so much coffee that the ship had given him his own mug and dubbed him "Coffee Man." I couldn't see how anyone could drink that much coffee in such hot weather, but he did.

"We've got good returns on the east–west lines," I replied, "and I've made all the waypoints with signal strengths. All you need to do is make a few north–south runs and we'll get it boxed in. We should be done in less than a day."

With that, I handed over the shift to Mark. I was eager to get some lunch and hit the sack. It had been a long night and day.

The *Catawba* turned slowly to the north, paying out cable again as we started boxing the pinger in. Once again, the TPL-21 went deep, with us paying out as much cable as we dared, leaving only three turns on the drum. We still didn't know how much cable we had out. All we could go by was the depth sensor in the towfish and a rule of thumb for deepwater searches, either a three-to-one or a two-and-a-half scope—that is, cable out versus water depth. It was good enough for what we were doing.

After scoffing down a hamburger and chips, I headed down into the pit, flashlight in hand, trying not to make any noise because people were always sleeping down there. A lukewarm shower made me feel a little better as I climbed up into my rack, slipping my tired body inside like a sardine into a can. I was using my sleeping bag this time to avoid using the ship's sheets because it was just easier. But after a week or two, it always began to stink. That just went with the territory. I tried resting my paperback book on my chest, but that was impossible because it hit the rack above me. I rolled over on my side, which was better, but then I couldn't see the pages. It was a waste of time and I decided to just pop a couple of sleeping pills and get some rest. Earplugs in, eyeshades on, and off to slumber. Even with the earplugs, I could hear the rumble of the ship's engines, shafts turning, propellers changing pitch, water pumps churning; it is never quiet on a T-ATF. Stacked in between the racks were numerous Pelican cases used by the Navy security team for who knows what.

As I drifted off to sleep, I silently chuckled at yesterday's events. One of the Navy security troops had left a massive piece of human excrement in a toilet and it got stuck to the side of the standard-issue stainless steel appliance. It was out of the water so after a while it stank up the berthing. With much reluctance, I called the chief engineer. The thing was huge, no doubt a result of the dietary supplements they all took to bulk up. The poor man came in, surveyed the situation, and dealt with it accordingly, getting it down the toilet. When he was done, I told him he deserved the Silver Star. He laughed and left us in peace.

Just as I was about to fall asleep, I heard a thump and saw a form fall past my bunk, then run off toward the hatch. It was Carl from the rack above me. He had slipped while climbing down and hit the deck hard. Finally I drifted off, wondering what the next day would hold for our team.

After a few hours of fitful sleep, I rolled out of my rack in the morning much like a butterfly extracting itself from a cocoon. There was just no room and no way to know who was on the deck when you got out. A portion of my belongings were stored underneath my mattress in a storage area. But as soon as I lifted my mattress and locked it into place, the bunk light was blocked. So, it was back to the flashlight because I couldn't turn on any lighting with others sleeping. I found my toilet kit and hit the shower so I could feel clean for a while in the blistering Arabian Sea sun.

The head smelled better after being decontaminated by the engineer, and it was off to the galley for bacon and scrambled eggs. I hit the bridge and discovered that Mark had done a good job during the day, managing to compete three reciprocal lines with good signal returns. I dragged out my laptop and began reviewing the WAV files. We had some really good returns, with Zilberman's pinger sounding out loud and clear from the depths. For all the Navy's money, we were using freeware, a program called Spectrogram 16, developed by an engineer working on the Seawolf program. As the waterfall display scrolled from left to right, the 37.5 kHz pinger was obvious. All we needed to do now was make sure we knew where it was. I examined the file's start time and correlated that with our navigation records (where the ship was when the file was created) and noted the peak signal time. From that, I was able to calculate where the ship and towfish were when the signal was the strongest. Weeks later, I would review the whole thing again for the salvage team and manage to pinpoint the location to within 150 meters, which was pretty good considering the water depth and possible errors with regard to knowing exactly where the towfish was at the time.

Soon after coming on shift, I notified the bridge that we were done and could start hauling in cable for the last time. Usually, during the last cable recovery after a job, you wash off the cable with fresh water and try to get decent wraps. But Frank, ever the perfectionist, was on deck taking far too long to get the job done. As a result, we were steaming outside of the area into unknown topography, and I didn't like it.

"Bridge, deck. Frank, can you hurry it up? I don't know how deep it is here and I don't want to hit bottom," I told him.

Frank replied with, "I'm trying to get some good outer wraps, but the level wind keeps messing things up."

"Well, get going. We're wasting time."

Eventually, we got all 6,000 meters of cable back on the drum, the towfish was on deck, and we were headed back to Bahrain. As we steamed away, longing for a hot shower, alcohol, and a quiet room, the Dukane pinger on Steve Zilberman's airplane ticked away, to die an honorable death in 336 hours.

On the way back to Bahrain, outside of the Strait of Hormuz, a couple of small vessels were heading right for us. This was not good because they could be carrying terrorists with rocket-propelled grenades, or the boats themselves could be packed with explosives. Within seconds, our security team on watch did what is called a "show of force." In other words, one guy manned the Browning M2 .50 caliber heavy machine gun while the other got ready with his M4 assault rifle fitted with the M204 grenade launcher. They didn't point their weapons at the vessels but made it clear they were ready to defend our ship, with lethal force if needed. They also had the M79 grenade launchers on board, though they claimed they only carried parachute flare rounds. *Right.* We also had a long-range acoustic device that could transmit warnings in various languages and output an incredible level of very uncomfortable sound, designed to incapacitate anyone who was targeted.

Fortunately, both vessels wisely turned away. One was a small cargo ship loaded with cars being smuggled into Iran and the other a cabin cruiser that was probably his security detail. I wonder if they ever knew how close they came to getting "smoked."

By the time we were ready to leave Bahrain, our crew was hot, dirty, and spent. It was time to go home. We were tired of the call to prayer from the huge mosque across from the hotel, as well as the orange dusty haze that always seemed to hang over the city from nearby sandstorms. Fortunately, we were able to get a quick flight home by first flying to Kuwait, then heading back to Dulles following a short layover.

You meet the most interesting people in the Kuwaiti airport waiting area, ranging from soldiers dressed in full kit, obvious journalists wearing their flak vests, and the occasional female covered from head to toe in a burka. Of course, we had seen that before in Manama, where there were usually four or five women dressed in the same manner, all dutifully following their husband.

It was when we landed at Dulles Airport that strange things started to happen. The date was May 3, 2011. The flight was packed with military and journalist types, and as soon as it was permitted, every cell phone on the airplane simply lit up—dozens of phones ringing all at once. As we slowly walked off the jet, I could hear a multitude of huddled voices whispering about something—excited voices, while dripping with emotion, interspersed with a few chuckles or somber laughs. *What in the hell is going on?* I wondered.

On the way to immigration, I cornered a Navy officer and asked him what was happening.

He glanced around, then said, "They got bin Laden!"

"What?"

"THEY GOT BIN LADEN!"

I stood, dumbstruck for a couple of seconds, then asked, "Is he alive?"

"I don't know . . . But they got him."

The man walked off down the hallway, leaving me in disbelief. Then I remembered something that had happened on the USS *Catawba*. Before our arrival in Bahrain, we took on over 150 pallets of hazardous waste in 55-gallon steel drums from various US Navy ships—stacked three high on our fantail in Dubai, Saudi Arabia. I remember walking around the deck looking at the names of all the ships. Many of the drums were from the USS *Carl Vinson*.

Later news accounts said that after the raid, reports at the time stated that US forces had taken bin Laden's body to Afghanistan for positive identification, then had him buried at sea, in accordance with Islamic law, within twenty-four hours of his death. Subsequent reporting has called this account into question, citing, for example, the absence of evidence that there was an Imam on board the USS *Carl Vinson*, where the burial was said to have taken place.

I thought about the perfectly sealed steel drums on the *Catawba*. *Nah, they wouldn't do that . . . would they?*

14

2015: The SS *El Faro*

A lot of things happened to me in the years following the search for *Bluetail 601*, some good, but mostly bad. Since 1994 I had been married a woman who will remain unidentified for a host of reasons. Because we were both older, the marriage did not produce any children, and the first few years were reasonably happy. But this was a first marriage for both of us, and living under the same roof is where you really get to know a person.

Over time, as I came home from Navy jobs, I began to notice stacks of wine and beer bottles in the recycling bin. Clearly, there was a lot of partying going on while I was out at sea. It all came to a head one Halloween night when I discovered numerous love letters and cards between her and another man. I hired a private investigator, and my suspicions were proven correct. It was then that I knew our marriage was done.

We sold our house, she went to another state to be with family, and I moved into a townhouse in Maryland. Eventually, she sued for divorce, and instead of settling, I elected to go to court because I really could not afford alimony on Phoenix International's shop wages. That was a big mistake. However, at the time I could not imagine that any court would award this woman alimony given her behavior. I imagined wrong, and it eventually cost me a considerable sum to settle the case.

During the years this court case was going on, I fell into a deep depression from the stress of it all. I cannot speak about what depression is like for others, but for me, it felt like my body was a tightly wound spring that never relaxed. I couldn't sleep, drank too much, and felt as though I was headed into an abyss as I watched myself drift slowly, but surely, toward bankruptcy. Due to the actions of her lawyer, I was forced to keep large amounts of cash in my home because my bank account had already been frozen once, while I was offshore on a job, of course. It didn't help that I eventually spent thirty-two days at sea nursing a hernia. All of this happened during the search for the SS *El Faro*, and I was faced with no way to pay my bills while being stuck on a Navy ship far from land. But then, there was still the *El Faro*, one of the few US registered ships to be lost at sea in a long time.

On September 30, 2015, Capt. Michael Davidson stared through the bridge windows and probably didn't like what he was seeing. While he spoke, his words were recorded by the ship's voyage data recorder bolted to the top of the bridge behind the main mast:

Uh—the word is pretty much what we knew this morning. we're—we're—we're on this intercepting and we've diverting to the south a little kind of take it wide and stay out of the—the deeper swells. and uh we—we could we're predicting uhh forty knot winds umm on our starboard beam. tonight. no—no—it's on the stern now thirty. (oh boy/or forty) we're looking at fifty knots on the starboard beam. the latest and greatest.—but that's not hurricane force.

The winds may not have been hurricane force that day at 3:00 p.m., but they were going to eventually sink his ship and kill everyone on board in less than seventeen hours. Davidson wasn't particularly liked by all his former shipmates, some of whom described him as "magnanimous but arrogant, with little interest in the details of running a ship." Seaman Kurt Bruer, who had worked under Davidson, called him "one of the laziest captains I'd ever sailed with," saying that Davidson spent most of his time in his cabin instead of walking the ship as other captains would do. Of course, it's easy to trash a former boss when he's not around to defend himself. Previously, Davidson had taken a 160-mile detour to avoid Tropical Storm Erika, although that extra time and fuel may have made him look bad in the eyes of management.

Others praised him, such as Nick Mavodones, who had sailed with David-son before: "Any time you're a mariner, you want to exercise prudence oper-ating a vessel, and I'm sure he was doing that earlier this week when they got underway from Jacksonville. He's been sailing these ships for many years and is very experienced and, I think, a very good sailor."

People who have borne the responsibility of captaining a ship say David-son's failure to deal with Hurricane Joaquin could have happened to any ship's master. "He seemed like a pretty normal captain," said George Collazo, a Seattle-based ship captain. "He could have done the same thing a hundred times and been fine."

As with most things, the truth was probably somewhere in between. Even so, Davidson was a professional mariner and made a good living, right up un-til the time he went down with the *El Faro*.

Location: 36 Miles NE of Crooked Islands, Bahamas
Date: October 1, 2015
Vessel: *El Faro*, Registration IMO: 7395351
Operator: TOTE Services
NTSB Number: DCA16MM001

On Thursday, October 1, 2015, about 07:15 a.m. eastern daylight time, the US Coast Guard received distress alerts from the 737-foot-long roll-on/roll-off cargo ship El Faro. The US flagged ship, owned by Sea Star Line, LLC, and op-erated by TOTE Services (TOTE), was 36 nautical miles northeast of Acklins and Crooked Islands, Bahamas, and close to the eye of Hurricane Joaquin. The ship was en route from Jacksonville, Florida, to San Juan, Puerto Rico, with a cargo of containers and vehicles. Just minutes before the distress alerts, the El Faro master called TOTE's designated person ashore and reported that the ship was experiencing some flooding. He said the crew had controlled the in-gress of water, but the ship was listing 15 degrees and had lost propulsion. The Coast Guard and TOTE were unable to reestablish communication with the ship. Twenty-eight US crewmembers and five Polish workers were on board.

In the transcript of the bridge communications from the voyage data re-corder, excerpted on the pages that follow, these individuals were identified by the National Transportation Safety Board (NTSB):

2M Second Mate—*El Faro*

3M Third Mate—*El Faro*

3AEX Unidentified Third Assistant Engineer (3AE1, 3AE2 or 3AE3)—*El Faro*

AB-1 Able Seaman 1—Helmsman 1—*El Faro*

AB-2 Able Seaman 2—Helmsman 2—*El Faro*

AB-3 Able Seaman 3—Helmsman 3—*El Faro*

CAPT Captain—*El Faro*

CM Chief Mate—*El Faro*

SUP-1 Alternate Chief Engineer of *El Faro* acting as a supernumerary—*El Faro*

USCG Transmission from the United States Coast Guard

? Voice unidentified; an unknown speaker

(All bracketed comments are part of the transcript. Asterisks indicate unintelligible words.)

September 30, 2015 — 17 hours before sinking
USCG-VHF 14:38:56.8–14:39:46.7

sécurité. sécurité. sécurité. all stations. all stations. all stations. this is the United States Coast Guard channel sixteen. the National Hurricane Center has issued a hurricane warning for the central Bahamas including Cat Island—(the Eleutheras)—Long Island—Rum Cay and San Salvador. The National Hurricane Center has issued a hurricane watch for north(west) Bahamas—including Abacos—(Berry) Island—Bimini—Eleuthera—Grand Bahama Island—and (new providence). The Coast Guard requests all mariners * extreme caution * *. the Coast Guard is standing by on channel sixteen. out.

This was a standard warning to all mariners transmitted by the US Coast Guard, in this case advising that the USCG was standing by on VHF channel 16. It was a definite warning to the *El Faro*'s crew, though there was only so much they could do about it. They couldn't outrun the hurricane, but only divert their course as much as possible given the prevailing winds and their best speed. The date was September 30, 2015, and the clock was ticking for the SS *El Faro*.

The heading of the ship was all-important because that determined how severe the rolls were going to be and ultimately their speed made good (the adjusted speed; generally takes into account wind and current, which affect the actual speed through the water). They didn't want 50 knots on the starboard beam.

CAPT 16:16:01.9–16:16:04.6

oh. no no no. we're not gunna turn around—we're not gunna turn around. [there is a brief unintelligible comment by either CM or AB-1 during the time the CAPT is speaking.]

CAPT 16:16:06.9–16:16:11.1

the—the—the storm is very unpredictable—very unpredictable. Turning around in the prevailing seas would have been almost suicide. In other words, they were stuck in the situation.

CAPT 16:16:24.6–16:16:47.6

so we get away from all that—when we leave San Juan—I wanna come home Old Bahama Channel. and not even get tangled up with this thing—that's continuing to come up the islands just—just off the Florida coast gets in the Gulf Stream there and shoots up outta here—it's just how do you deal with it? just through the Bahama channel. now uh—I asked the office.

AB-1 16:16:48.5–16:16:49.7

any word back yet?

CAPT 16:16:50.1–16:16:55.9

no. cause it's a hundred and sixty more miles. that's more fuel. you know?

Unfortunately for Davidson, economics played a hand in his decision-making process. A large ship like the *El Faro* makes money by using as little fuel as possible. A large diversion around the storm would add distance to their route and result in more fuel usage. But there comes a time when all that goes out the window because, ultimately, they would be fighting for their survival.

September 30, 2015 — 8½ hours before sinking

ET 23:05:08.7–23:05:10.4

hey captain sorry to wake ya.

3M-ET 23:05:12.0–23:05:20.6

naw—nothin' and uh the latest weather just came in.—and umm—thought you might wanna take a look at it.

3M-ET 23:05:43.1–23:06:43.4

uhh well it's—the—the—the current forecast has it uhh—max winds um a hundred miles—an hour. at the center.—umm and if I'm lookin' at this right—um—and it's moving at—at two-three-zero at uh five knots. so I assume it stays on that same—moves that same direction for say the next five hours. and uh so it's advancing toward our trackline...

AB-3 23:50:39.1–23:50:41.1

do you know where your EPIRB is?

The EPIRB that able-bodied Seaman 3 is talking about is an emergency position indicating radio beacon. They all have serial numbers that can be tracked back to the ship and communicate directly to a search and rescue satellite. Most ships carry several of them, with some of them designed to float free if the ship sinks. All of the inflatable life rafts carry them as well.

AB-3 23:50:41.6–23:50:44.4

wish I had one like uh @A3M had one. one of those little pocket ones.

AB-2 23:50:45.2–23:50:45.7

yeah.

October 1, 2015 — 6½ hours before sinking

At this point in the voyage, the *El Faro* began to take heavy rolls. In a ship of this size, that makes it difficult to stand up on the bridge and do your job. Anyone on the ship's bridge would need one hand on a rail at a minimum. It is also very fatiguing because a lot of your physical strength is being used up just staying in one place.

AB-2 01:18:26.0–01:18:29.3

* * *. * one roll we took.

AB-2 01:18:32.7–01:18:35.5

biggest one since I've been up here.

2M 01:19:41.5–01:19:43.7

nope. hurry up and get the hell outta here.

AB-2 01:19:45.2–01:19:50.0

can't pound your way through them waves—break the ship in half.

2M 01:40:04.5–01:40:33.7

[sound of quick laugh] usually people don't take the whole umm—uh—survival suit—safety meeting thing very seriously. then it's "yeah—whatever. it fits" but nobody actually seems to see if their survival suit fits. I think today would be a good day [sound of laugh] for—for—for the fire and boat drill—just be like—"so we just wanna make sure everyone's survival suit fits" and then with the storm people are gunna (go/be like) "holy #. I really need to see if my survival suit fits—for reaaal." [laughter throughout]

2M 01:40:34.5–01:40:37.6

nobody ever takes these—the drills—seriously.

The survival suits the second mate is talking about are special insulated suits that protect persons in the water from the cold of the ocean. Depending upon the type and style, they are very difficult to don, and during fire and lifeboat drills, one person is always selected out of the crew to put one on and be timed while doing it. We used to call them "Gumby suits" because that's what they make the user look like. The reality is that while the suits do offer protection, they won't keep you alive forever, only extend the time you can survive in the water. If the water is really cold, you might last a few hours.

October 1, 2015 — 4½ hours before sinking

2M 02:53:48.7–02:53:50.8

she's righting herself.

M1 02:53:53.4–02:54:01.4

[sound of electronic tone consistent with steering stand alarm.]

AB-2 02:53:55.3–02:53:56.2

hell's that?

2M 02:53:58.5–02:54:00.3

she off course?

AB-2 02:53:59.9–02:54:00.6

she is.

AB-2 03:22:37.8–03:22:39.0

hear that wind out there?

2M 03:22:42.5–03:22:44.5

yup we're gettin' into it now.

2M 03:22:46.5–03:22:47.9
hello Joaquin.
CM-ET 04:36:47.1–04:36:49.6
container where? the second deck? *.
CM-ET 04:37:06.3–04:37:07.1
trailer.
CM-ET 04:37:08.5–04:37:09.1
leaning over.

The above transcription indicates that the crew was losing the helm, as in their ability to control the ship's heading. It could have been a control failure or simple loss of propulsion. In the existing weather conditions, this is a very serious matter, one that could determine whether or not they lose the ship.

October 1, 2015 — 3 hours before sinking

CM-ET 04:37:27.7–04:37:29.1
yeah we're heelin' over *.
CM-ET 04:37:34.1–04:37:36.2
I'll pass it on to the captain.
SUP-1? 05:10:51.1–05:10:53.0
what's the wind speed?
CAPT 05:10:52.6–05:10:57.4
we don't know. we don't have (any) anemometer.
CAPT-ET 05:43:21.0–05:43:22.5
bridge—captain.
CAPT 05:43:28.3–05:43:31.7
* * *.
CAPT-ET 05:43:36.4–05:43:38.5
we (got) a prrroooblem.

Losing the wind speed indicator in such conditions is not uncommon. The worst part is that if they have no idea what the wind speed is, the crew is unable to discern the seriousness of the situation. However, anyone who has spent a lot of time at sea can usually estimate the wind speed by the sound and what they see on deck. But if it is nighttime, you can't see the direction of the seas without shining a spotlight on the water. Knowing the direction of the swells is all-important because it determines what heading you should be holding. With the helm control out, the *El Faro* was left at the mercy of the wind and seas, which can be catastrophic.

October 1, 2015 — 2 hours before sinking

CAPT 05:43:47.5–05:44:02.3

watch your step—go down to three hold—go down to three hold. * down there * start the pumping right now * (probably just) water * * *.

CM 05:43:58.6–05:44:06.2

* * * suspected leak (in) (he said) * *.

October 1, 2015 — 1½ hours before sinking

CAPT-ET 05:52:38.6–05:52:52.5

okay what I'm gunna do—I'm gunna turn the ship and get the wind (on the north side) right there and get (it going) more (in that) direction get everything on the starboard side give us a port list and (um see) if we'll have a better look at it.

So now it appears they do have helm control. Did they bring it back online after initially failing?

CAPT-ET 05:52:58.3–05:53:00.0

yeah—so it is the scuttle?

CAPT-UHF 05:54:44.1–05:55:00.1

let me uhh—I just got off the house phone with the chief. I want to turn the uhh—ship to port—get the list on the port side there. can you access it right now? how much water is there?

CM-UHF 05:55:00.4–05:55:08.9

(ya got) water against the side just enough to (go/throw/pour) over the edge of scuttle about knee deep (in here) water (rolls) right over.

CAPT 05:59:37.4–05:59:54.2

a scuttle popped open and there's a little bit of water on in three hold. they're pumping it out right now. the mate's down there with @SUP-1 he's down (in the/he's closing / * *) the scuttle.

The scuttles are steel flaps on the main deck where most of the tractor trailers are stored. They control the amount of water that can come in from the outside of the ship. If one of them has failed and is stuck open, then that whole area will flood under the prevailing sea conditions.

October 1, 2015 — 1 hour before sinking

CAPT 06:45:22.1–06:45:28.6

* * (a lot of) water in the cargo hold area.

2M 06:45:28.2–06:45:32.9

(well then) suck out the water.

CAPT-ET? 06:54:33.3–06:55:10.5

(it's) miserable right now. we got all the uhh—all the wind on the starboard side here. now a scuttle was left open or popped open or whatever so we got some flooding down in three hold—a significant amount. umm everybody's safe right now we're not gunna abandon ship—we're gunna stay with the ship. we are in dire straits right now. okay I'm gunna call the office and tell 'em * *. okay? umm there's no need to ring the general alarm yet—we're not abandoning ship. the engineers are tryin' to get the plant back. so we're workin' on it—okay?

October 1, 2015 — 40 minutes before sinking

CAPT-FBB 07:01:04.9–07:01:13.1

this is a marine emergency. yes this is a—ah marine emergency and I am tryin' to uh also notify a Q-I.

CAPT-FBB 07:05:59.6–07:06:20.5

I have a marine emergency and I would like to speak with a Q-I. we had a hull breach—a scuttle blew open during a storm—we have water down in three hold—we have a heavy list—we've lost the main propulsion unit—the engineers cannot get it goin'. can I speak with a Q-I please?

CAPT-FBB 07:07:15.4–07:07:59.6

yea. I'm real good. we have uhh—secured the source of the water coming into the vessel. uh a scuttle—was blown open uh—by the force of the water . . . however—uh—three hold's got considerable amount of water in it. uh we have a very—very—healthy port list. the engineers cannot get lube oil pressure on the plant therefore we've got no main engine. and let me give you um a latitude and longitude. I just wanted to give you a heads up before I push that—push that button.

The above conversation (only one side presented) is between the captain and TOTE's designated person ashore, whom the captain is advising that the crew is in trouble and considering abandoning ship. The captain has yet to "push the button," as in declare a general alarm. It is amazing that he was even able to communicate with his office under these conditions.

October 1, 2015 — 30 minutes before sinking

CAPT 07:13:55.3–07:13:57.3

wake everybody up. wake 'em up. [exclaimed in an urgent, almost angry tone.]

CAPT 07:13:60.0–07:14:03.3

we're gunna be good. we're gunna make it right here * *.

2M 07:14:22.3–07:14:25.1

he's just tellin' us the same thing. he can't do anything with this list.

? 07:17:51.4–07:17:54.3

Consider ourselves * *.

? 07:17:52.9–07:17:56.4

* no R-P-M * * we can't do anything.

The captain and crew of the *El Faro* have finally realized that the ship is beyond saving. As experienced seamen, every one of them knows that their chances of survival are slim to none and they are behaving like any person would under the circumstances. They must be terrified of dying but are trying to remember their survival training, which will be of limited benefit in the end.

October 1, 2015 — 13 minutes before sinking

CAPT-UHF 07:26:27.9–07:26:35.4

hey mate chief mate. (this is/just a) heads up I'm gunna ring the general alarm. get ya muster while down there. muster all mate.

CAPT-ET 07:26:49.7–07:27:02.5

alright captain here. just want to let you know I am going to ring the general alarm. I am going to ring the general alarm. (you don't have to/ya know we're not gunna) abandon ship or anything just yet. alright we're gunna stay with it. is the chief there?

?-UHF 07:28:32.1–07:28:34.8

cap'n you gettin' ready to abandon ship? * *.

CAPT-UHF 07:28:37.8–07:28:49.4

yeah what I'd like to make sure everybody has is their immersion suits and uh—stand by. get a good head count. good head count.

The above exchange does not make sense. They're either abandoning ship or they are not. The only explanation is that the captain has not fully accepted the fact of what they have to do.

October 1, 2015 — 10 minutes before sinking
CAPT 07:29:28.8–07:29:32.3
alright let's go ahead and ring it—ring the abandon ship.
M1/M2 07:29:33.3–07:29:40.0
[sound of a high frequency bell ringing in seven pulse tones.]
M1/M2 07:29:40.0–07:29:47.9
[continuous sounding of a high frequency bell for about eight seconds.]
CAPT 07:29:53.1–07:29:54.9
tell 'em we're goin' in. [spoken in a raised voice]
CAPT 07:30:03.3–07:30:05.3
okay buddy relax. go 'head second mate.

October 1, 2015 — 9 minutes before sinking
CAPT 07:30:16.3–07:30:17.4
bow is down.
CAPT-UHF 07:31:03.5–07:31:08.4
yeah—yeah—yeah—get into your get into your rafts * * throw all your rafts (in/to) the water. [yelled throughout]
?-UHF 07:31:10.9–07:31:13.1
throw the rafts in the water—roger.

The *El Faro* has started the process of going under by the bow. There is no longer any doubt in anyone's mind about what is going to happen.

October 1, 2015 — 8 minutes before sinking
CAPT 07:31:53.3–07:32:00.8
come on @AB-1. gotta move. (we) gotta move. you gotta get up. you gotta snap out of it—and we gotta get out.

AB-1 07:32:01.1–07:32:01.5

okay.

CAPT 07:37:45.1–07:37:46.7

hey. [yelled] don't panic. [spoken loudly]

October 1, 2015 — 2 minutes before sinking

CAPT 07:37:48.7–07:37:50.7

Where are the life preservers on the bridge? [yelled]

AB-1 07:38:14.0–07:38:15.2

you gunna leave me.

CAPT 07:38:15.9–07:38:17.3

I'm not leavin' you let's go. [spoken loudly]

AB-1 07:38:21.5–07:38:22.5

[sound of scream]

October 1, 2015 — 1 minute before sinking

AB-1 07:38:35.7–07:38:38.2

(I need someone to/help somebody) help me. (you don't wanna/you gunna) help me? [yelled]

CAPT 07:38:39.0–07:38:40.0

I'm the only one here @AB-1.

AB-1 07:39:06.0–07:39:06.8

I can't. [spoken clearly and loudly]

AB-1 07:39:07.8–07:39:08.5

(I'm gone/I'm a goner). [spoken loudly]

AB-1 07:39:14.1–07:39:15.1

just help me. [yelled]

CAPT 07:39:15.3–07:39:16.6

@AB-1. [exclaimed] let's go. [spoken loudly]

SS El Faro *goes down by the bow*

M1 07:39:32.2 10/1/2015

[sound of building low frequency rumble until end of recording.]

CAPT 07:39:38.0–07:39:38.4

@AB-1. [yelled]

CAPT 07:39:39.0 10/1/2015
it's time to come this way. [yelled]
? 07:39:41.7–07:39:41.8
*—[yelling cut off by the termination of the recording.]
ALL CHAN.
07:39:41.8
END OF TRANSCRIPT.
END OF RECORDING.

Personally, I found the voyage data recorder transcript from the *El Faro* difficult to read. It is like watching a movie you've seen before where everyone dies in the end. You know what's coming. Trying to get into those survival suits is difficult at best on a good day. Trying to get into one of those things during a hurricane while the ship is sinking—almost impossible.

Another problem is getting into a raft if you do get off the ship. After attending many water survival courses over the decades, I can say from firsthand experience that trying to get into an inflatable raft while wearing an immersion suit is difficult, even under perfect conditions. I cannot fathom what it would be like in the open ocean.

But survival is a great motivator, for all the good those suits did in the end. During fire and lifeboat drills, we always joked that they only made it easier for the Coast Guard to find your lifeless body.

When the *El Faro* began her plunge to the bottom, several things happened, though in what order is difficult to discern. Certainly, the ship's main mast was torn away either by the wind or a massive wave, taking the voyage data recorder with it. Later inspection showed that when the mast was ripped off, it took a 2- or 3-foot-square chunk of the deck with it, which also held the data recorder. Either at that time or shortly thereafter, the top two decks of the bridge were neatly sliced off the ship, probably by a massive wall of water. The break was extremely smooth and accurate, as if done with a massive cutting torch.

It is impossible to say how many of the crew made it to their muster points; given that only nine minutes elapsed between the general alarm and the sinking, it is doubtful they all made it out of the ship's interior. In fact, it is unlikely that many of the crew even had the time to put on their immersion suits because that is typically done at the muster point. However, at least one crew member was able to don his, because he was sighted during the initial surface search, presumably floating dead on the surface.

Of the more than 350 shipping containers carried by the *El Faro*, only 2 were still on deck when we surveyed the ship with CURV 21. One of them was a liquid product tank, crushed by the giant fist of the water pressure at depth.

While the ship did carry numerous self-inflatable life rafts, it would have been very difficult to get inside of them given the sea conditions. The two open lifeboats would have been of little use as well. Overall, despite all the precautions taken by trying to avoid the hurricane, the *El Faro* was consumed by the storm. The ship was doomed. It was too heavily loaded, too old, and not fast enough to get out of the way of Hurricane Joaquin.

As with most black boxes, the one on the bridge of the *El Faro* was fitted with an acoustic beacon that could be detected with a pinger locating system. Unfortunately, when we started our search for the sunken ship using the Navy's TPL-25 system, we didn't hear a thing, even though we felt confident we were in the right area. As with most devices, these beacons are not always maintained on a regular basis. In fact, Malaysian Airlines was not even sure what the frequency was of the beacons attached to the flight data recorders on Flight 370. Therefore, it was easy to suspect that the *El Faro*'s voyage data recorder beacon batteries had never been changed; they typically last about five years.

When we completed loading all our search gear onto the USNS *Apache*, a T-ATF fleet tug, we had only a general idea as to where to search for the missing ship. The NTSB had supplied us with the last GPS fix from the ship's automatic identification tracking system, or so we thought. Also, using our connections with the National Oceanic and Atmospheric Administration, we had a satellite image showing an oil slick, undoubtedly from the *El Faro*. That location was where we would center our 5-by-10-nautical-mile search area.

This would be a multiphase approach; we would start off with my TPL-25 system to see if we could detect the acoustic beacon on the voyage data recorder because this system had a far better detection range than a towed sonar. If nothing was heard from the pinger, we'd switch to side-scan sonar, followed up by a target identification dive with CURV 21. The area was deep at over 15,000 feet, but we'd all been there before. It was no big deal.

The big issue with any towed search system is the amount of time it takes to turn everything around at the end of each search line. There was no getting around this as the Navy did not have a deep autonomous underwater vehicle system at the time. The problem came when people tried to cut corners during the turns by dumping cable at the beginning of new lines. The tow body used on my pinger locator needed about 1 knot of speed through the water column to maintain stability. Unfortunately, the night shift paid out cable far too fast, resulting in the towfish spinning on the end of the cable like a top. What made me angry was that they could see this happening from the pitch and roll indicator on the topside receiver but either ignored or didn't notice it. Consequently, all the rigging connected to the towfish became fouled, creating considerable noise from the turbulence.

After several days of searching with the pinger locator, we heard nothing. So, down went the Orion, a 6,000-meter-rated low frequency towed side-scan sonar. Using this vehicle would be even slower than using the TPL because the search lines were closer together.

As the search dragged on, one morning while taking a shower in the *Apache*'s shower stall, I felt a small lump in my groin. Fortunately, because we expected to recover human remains on the job, we had a corpsman on board. It only took a few moments before he diagnosed me with an inguinal hernia. Since we were stuck at sea, there was little to do about it. All I told him was that I didn't want to die out there, and he cautioned me to keep an eye on it. For me to get off the ship would delay the search several days and cost over $400,000. I wasn't ready to do that and decided to stick it out. As a result, I spent thirty-two days at sea with a hernia.

As is usually the case on these operations, I was stuck on the night shift. About thirty minutes after I went off shift and ate lunch, the day shift found the *El Faro*. It was a massive rectangular target on the sonar's starboard channel. There was no mistaking it. Surprisingly, not long after we found the ship, our Navy client, Ric Sasse, discovered that the NTSB had not even given us the last GPS fix from the ship; the final position was buried in a long email thread. Even with that, we still found it. The next phase would involve sending CURV 21 deep into the ocean to survey the wreckage and find the voyage data recorder.

We launched CURV 21 without incident on November 1. Just before the dive, from our perspective, about five hours had been wasted while the ship cleared her stacks.

When any ship remains in one place running her engines at minimum speed, carbon deposits build up in her stacks, the exhaust systems for the diesel engines. If it gets bad enough, it can result in a stack fire, where the smokestacks literally catch on fire. Consequently, a ship needs to periodically clear her stacks, which means the vessel must steam at high speed for a while to clean things out. We had all been working at sea for decades and knew this was necessary. But the captain of the *Apache* seemed to do this all the time, whether or not it was needed. This was a waste of time for us because obviously we could not be diving if the ship was moving. But he was the captain and therefore could do anything he wanted, and all we could do was wait until he felt all was good.

We finally got CURV on her way to the bottom at 10:45 that morning. This first dive was aborted because we had no acoustic navigation. By the time I came on shift at midnight, the vehicle had already found the massive target identified by the Orion sonar, and there was no question it was the remains of the SS *El Faro*. A brief glimpse of the stern showed the ship's name in large letters: "EL FARO SAN JUAN." The stern was heavily down in the mud, indicating that this section of the ship had hit first. We couldn't even see the rudder or propeller.

However, our job was to find and recover the voyage data recorder, and we knew where it was supposed to be: on top of the bridge behind the main mast. But the top of the bridge was gone, along with two decks.

As we surveyed the ship, we noticed a lot of things—some expected, some not. The starboard lifeboat davits were still stowed, which showed that the crew never had a chance to deploy them. In addition, the life raft on that side was missing. Farther toward the bow, a massive tractor trailer was sticking out of the side of the ship like a huge metal tongue. One of the ship's funnels was nearly torn off and left hanging toward the stern.

What surprised us was that all of the ship's mooring lines were still attached to the deck and floating up vertically, creating an excellent ROV (remotely operated vehicle) trap and causing the ship to look like an upside-down spider web. We had to be very careful not to get CURV or our umbilical caught up in those things. The depth was 4,732 meters (about 15,520 feet)—deeper than the *Titanic*. Yet here we were, driving around this chunk of steel like it was nothing.

It soon became obvious that the voyage data recorder was not on the ship, which meant that now our task was to find the missing sections of the bridge that had been sliced off.

Our navigation at that depth was OK, but not great. We were using an ultra-short baseline navigation system, which meant that our positioning came from the surface via the pings between the topside transducer and the one on CURV. We did our navigation this way because we did not have time to put down an array of bottom transponders on our jobs. But while it made it much quicker to get operating, navigational accuracy suffered.

As it became obvious that we needed to search the debris field down current of the ship, we established numerous search lines that CURV would follow while using a scanning sonar to find the top sections of the bridge, and hopefully the voyage data recorder. After wasting much time embarking a *60 Minutes* film crew, we continued searching on the ship's west side, finding assorted junk, like an air-conditioning control panel, several shipping containers (one filled with lawnmowers), a white Toyota sedan, microwave ovens, the ship's gangway, and a tire-changing tool labeled "COATS RC-45."

Finally, at 7:10 a.m. on November 11, we located the top two sections of the bridge. They were sitting flat on the bottom with one of the doors open. Many windows were intact, but others were blown out, with electrical cables sticking through them. Unfortunately, after doing a thorough survey of the area, we still had not located the mast and voyage data recorder.

Our original plan was to deploy the XBot ROV from CURV to explore the interior of the bridge. XBot was a small battery-powered vehicle used previously by James Cameron to probe the interior of the *Titanic*. But the inside of the bridge was strewn with cables and the ceiling had collapsed, making the task impossible. Instead, we began to follow a long trail of plastic garbage bags to the north. We did find one section of tubular structure that we thought might be part of the mast but concluded that it had come from a different part of the ship. All we found was an area littered with dental retainers, appliances, electric fans, and cartons of milk.

Ultimately, Hurricane Joaquin was directly responsible for thirty-four deaths in the waters off the Bahamas and Haiti. Almost all of the deaths occurred when the US-flagged cargo ship *El Faro* was lost at sea near the Bahamas while Joaquin was moving through the area. The elusive voyage data recorder was finally discovered by a team from the Woods Hole Oceanographic Institution, using an autonomous underwater vehicle, the *Sentry*, fitted with cameras. The team literally photographed the entire debris field, something beyond our capabilities and funding using the Orion search vehicle. The US Navy and CURV 21 were ordered to return to the area and recover the recorder, which they did. The recorder provided considerable insight into what happened the day the ship went down.

15

2017: Shadow of the *Indianapolis*

F
ollowing the end of the SS *El Faro* operation, I started to hear rumors about another search for the USS *Indianapolis* (CA-35). The professional underwater field is very small, and projects in the works invariably leak out and make the rounds to various companies and people. I cannot remember exactly where I first heard about Paul Allen planning a search for the *Indy*—it may have been from the daughter of a survivor or a work associate—but eventually I tracked down the rumor and connected with a sonar expert, Gary Kozak, who had been hired to join the expedition to evaluate side-scan sonar records. Dr. Richard Hulver, a historian at the Naval History and Heritage Command, discovered a tantalizing clue, the identity of a landing ship tank, the USS *LST-779*, that was passed by the *Indianapolis* during her transit from Guam to the Philippines. This single fact allowed us to recalculate the path of the ship far more accurately, building on the research I had done for my 2000 expedition. Consequently, I dragged out my old records and started making new estimates of the ship's position when she was torpedoed.

Working with an old friend from Oceaneering International, Andy Sherrell, I studied this new information to see how it might better pinpoint the sinking location.

One of the intriguing clues I discovered was a photograph taken by a crew member of *LST-779*, Floyd Lambertson, showing the *Indianapolis* passing by their stern. The image suggested that the two ships were very close during the passing. Unfortunately, this image *could* have been taken in February, when both ships supported the invasion of Iwo Jima, or in July (the period of interest), when the cruiser passed the LST at 13:22 hours on the twenty-ninth. Therefore, both situations had to be considered: that the *Indianapolis* passed close by the LST during her transit from Guam, or that she was some undefined distance away where both ships were able to see and identify each other. The faded picture was definitely that of a Portland class heavy cruiser and even had the name of the *Indianapolis* scrawled on the back. Given the likely 50 mm lens used to take the image in 1945, I estimated the distance between the two ships at about ½ nautical mile. In addition, given that the LST was to the north of the *Indianapolis*, the bow of the cruiser was pointed in the right direction—west. That made it even more likely that this image was indeed taken during the Guam transit. As far as I knew, this was the last photograph taken of the *Indianapolis* before she was torpedoed.

Capt. Charles B. McVay's oral history (bracketed information included) states:

> We had no incidents whatsoever. We passed an LST [Landing Ship Tank] headed toward Leyte [Philippine Islands], as we were also, on Sunday, and talked to them. They were north of us and were they were [*sic*] preparing to go further north in order to get out of our area to do some anti-aircraft shooting. My instructions from Guam called for me to make an SOA [speed over-all] of 15.7 knots and to arrive at Leyte at 1100 Tuesday, 31 July.

I assumed the 13:22 position from *LST-779* to be the general area where that ship was when they communicated with and sighted the *Indianapolis*. To calculate the probable error in both the LST and *Indy* locations, I also assumed that the LST's location was accurate to within ±4 nautical miles. The value of 6 nautical miles was baselined as a reasonable *maximum* distance between the two ships, given that the horizon was about 9 nautical miles away (visibility was recorded as 8 nautical miles with 3/10 cloud cover). This resulted in an area of about 14 by 14 nautical miles, or an area of 196 square nautical miles.

Other information I studied was the various zigzag patterns used by the Navy during the war. It was a well-established fact that the *Indianapolis* ceased zigzagging at 2000 hours, or 8:00 p.m., on the evening before the attack. That allowed me to backtrack the course and estimate where in the pattern the *Indy* was located as she steamed past *LST-779* to the south. If I knew at what point in the pattern the ship was, then I could determine how far to the south the ship's base course was during the passing.

I then started digging through all the surviving logbooks from the *Indianapolis* to develop a history of what patterns the ship used and under what circumstances. It turned out that the *Indianapolis* used patterns developed by both the United States Fleet (USF) and the Royal Navy, a fact that shocked Dr. Hulver when I shared this information with him. He couldn't imagine a US Navy warship using a Royal Navy zigzag pattern. But that was clearly stated in the surviving logbooks. The question was, *Could a case be made for what pattern the ship actually used?*

What I discovered was that during the last four months in 1945 that the *Indianapolis* was at sea for which logbooks are available, the cruiser used a total of seven different zigzag patterns. Of these, four were USF patterns, while the remaining were designed by the Royal Admiralty.

Whether or not the zigzag maneuvers used were one- or two-hour patterns was important because, considering the timing of the encounter between the *Indianapolis* and *LST-779* and when they are known to have ceased zigzagging, the use of a one-hour pattern suggests that the ship was either just finishing or just beginning a pattern. If a two-hour pattern was being used, then the cruiser would have been approximately halfway through the pattern.

Capital ships like the USS *Indianapolis* do not typically transit unescorted, which makes the circumstances of the sinking unusual because the cruiser was not fitted with any sound-detecting equipment. That task was usually performed by escorting destroyers.

During the months ranging from January to early May 1945, the *Indianapolis* was either escorting military transports, part of a task force, or being escorted by one or more destroyers. During that period, there were only two situations in which the *Indy* was steaming unescorted: during a transit from Iwo Jima to Guam on March 7 and while sailing to Mare Island (California, for her last refit) from Pearl Harbor in late April.

During the transit to Guam, the *Indianapolis* used USF plan no. 8, which is a fast one-hour pattern in that the ship was doing turns for 20 knots. While sailing to Mare Island, the cruiser used Admiralty pattern 40 Zebra, which is a slower two-hour pattern whereby the cruiser maneuvered to port of her base course at 16 knots; a mirror image of the same pattern is identified as 19 Zebra in the Admiralty documentation. When the *Indianapolis* used that pattern in late April, she was steaming at a speed that would make sense from the standpoint of her assigned speed of advance during her final transit from Guam to the Philippines in late July of 1945. If the cruiser was doing turns for 16 knots while using 40 Zebra, her speed made good would have been 14.7 knots, which is the zigzag speed used in these studies.

Given that the July 1945 logbook was lost with the ship, it is impossible to know what pattern the USS *Indianapolis* used during her final transit. (Cmdr. Janney, the navigator, did not survive.)

I eventually developed two scenarios: one where the *Indianapolis* was on base course (using a USF pattern) when the ship passed *LST-779* to the south, and the other where the ship was using the Admiralty 40 Zebra pattern and was about 4.5 nautical miles to the south during the encounter. Regardless of which pattern the ship was using, both placed the *Indy* over 30 miles west of the official Navy sinking location, assuming the ship ceased zigzagging at 2000 hours (which was well established) and had an over speed of advance of 15.7 knots while in pattern (using revolutions for about 17 knots to make up for the loss of speed while zigzagging). These placed the ship both on track and possibly to the south, depending upon the actual course.

The truth of the matter was that it was all an educated guess because we didn't even know the pedigree of the Navy's official sinking location. All we knew was that it was referenced during the initial dispatch sent by Capt. McVay from the USS *Ringness* informing the Navy of the loss of the ship. I felt that it was an estimate calculated by McVay after the fact and based upon where he *thought* they were when they were attacked. It's also possible that this was the location calculated by Cmdr. Janney after they were torpedoed and was the one sent out on the distress calls no one heard. But that meant that McVay would have had to commit the position to memory and retain it for the five days he was in the water, and that didn't seem plausible to me.

The reader must understand that figuring out the actual sinking location of the *Indianapolis* required guesswork because there were so many variables involved. The key objective was to get the location close enough to be able to define a search area.

I also decided that if the account of the encounter between the USS *Indianapolis* and USS *LST-779* was as suggested, then it would appear to be impossible that the *Indy* was attacked as far west as Navy pilot Chuck Gwinn's sighting of 150 survivors in the water; therefore, it must have been to the east. It is also unlikely that the ship sank where the Navy said it did. Considering the longitude of the new calculated sinking location, and assuming the accuracy of the Gwinn location, the survivors would have drifted to the west at a speed of roughly .32 knots.

The location identified by Lt. Cmdr. Hashimoto was well to the north of the *Indy*'s track and could not be correct, if the ship was reasonably on course as suggested by Capt. McVay. While it agreed with the new calculated position in longitude, it was considerably to the north by 36 nautical miles. There was no other evidence to suggest that the *Indianapolis* was in that area; in reality, the *LST-779* sighting and the fact that this ship steamed past the sinking location to the north without sighting any wreckage or survivors indicates that the ship was somewhat south of the Peddie route. Hashimoto's *I-58* was positioned prior to the attack at a known intersection of two routes: the Peddie and one running from Palau to the north. If the *Indianapolis* had not been generally where she was supposed to be, the ship never would have been sunk. Unfortunately, little survives of any Japanese records of the *I-58* (and the *Indianapolis* torpedoing), other than a hand-drawn maneuvering chart created during McVay's court-martial hearing and Hashimoto's book describing Japanese submarine operations during WWII.

By mid-2017, I felt I had developed a pretty good theory of where the *Indianapolis* sank; now it was time to see if a return to the Philippine Sea could be funded. Andy and I had been discussing a search for the ship with the Woods Hole Oceanographic Institution for some time because they had a Remus 6000 vehicle that could do the search.

In the meantime, I was guided to the producers of a new dramatic film about the sinking by Maria Bullard, head of the Second Watch Organization, an offshoot of the USS *Indianapolis* Survivors Organization. In short order, I was having conference calls with Richard Rionda Del Castro, one of the film's producers, as well as Tim Cavanaugh, another producer and actor. The film was titled *USS Indianapolis: Men of Courage* and was slated to be directed by Mario Van Peebles. I contacted Woods Hole and managed to put together an operational plan and budget for what it would take to find the ship. The producers thought an actual search for the ship might help promote their film.

The people associated with the *Indianapolis* film were extremely nice to me, inviting me down to the set in Mobile while they were shooting scenes on the USS *Alabama*, which was sitting in as the *Indianapolis*. I also started a friendship with actor Yutaka Takeuchi, who played Capt. Hashimoto in the film. Yutaka had contacted me to learn more about Hashimoto, and I sent him some documents from Hashimoto's testimony during the McVay court-martial. (Yutaka is a great guy and turned in a stellar performance in the film. We are still friends to this day.)

I went down to Mobile to hang out with Yutaka and watch them shoot portions of the film. It was a lot of fun and made me respect how much work any film production is. The amount of equipment they needed on set was incredible and involved scores of very talented people. It was extremely hot down there and I couldn't imagine wearing all those heavy WWII-period uniforms. I usually ended up huddled under the number two gun turret of the *Alabama*, trying to stay hydrated and out of the sun.

In the end, they were unable to fund the search due to the $2 million price tag. That was just what it would cost to have a realistic chance at finding the *Indy*. In fact, any search for the *Indianapolis* was going to be expensive due to the remote nature of the area. The sinking location was in the middle of the Philippine Sea, between Guam and the Philippines, and there were no suitable support ships in Guam. Consequently, they would have to come from Indonesia, and that would mean transit costs. It would take someone with very deep pockets to find this ship. Little did I know that there was such a person. His name was Paul Allen.

Allen was well-known as one of cofounders of technology giant Microsoft, along with Bill Gates. He had amassed an immense fortune of probably close to $20 billion and owned the largest private yacht in the world, the *Octopus*, a 414-foot-long vessel outfitted with the latest in underwater technology, including a Bluefin-9 AUV (autonomous underwater vehicle), an ROV (remotely operated vehicle), and a small manned submersible that his guests could use to explore the ocean floor. I think there was also a small helicopter on board, as well as a smaller boat that could be launched from the mother ship. I didn't know it at the time, but Allen also had a keen interest in WWII history and had been using the *Octopus* to explore various areas of the world where ships had been sunk during the war.

One thing I learned was that having that much money created baggage. Certainly, everyone wants to be rich, but I heard from people who knew Allen that he literally couldn't get out of bed in the morning without being sued by someone. The problem is that when you have that much money, there are a lot of people trying to steal it from you. But I have to say that I admired what he did with what he had through his company Vulcan Inc.

Paul Allen was very secretive about his business dealings, but during his lifetime he invested heavily in his hometown of Seattle, saved numerous antique aircraft via his Flying Heritage & Combat Armor Museum, funded brain research, and helped preserve the integrity of the oceans. Allen was already famous, so he didn't need fame. But apparently one of the things he wanted was to see the *Indianapolis* found.

The *Indianapolis* was not an easy target; however, from the standpoint of WWII history, it was the *only* target. The problem with the *Indianapolis* was that no one we knew of had enough money to do a proper search that had a high probability of finding the ship. But Allen did, and it was not long before he made a commitment to finding her, if it was technically feasible. It was. He already had a quasi-research ship, the *Octopus*. But Allen had difficulty using it as a serious platform for his interest in finding WWII ships due to his extensive social calendar. I imagine there were endless streams of celebrities wanting to spend time on that ship. Who wouldn't?

By the time I became involved with Vulcan, Allen had already located and surveyed the *Musashi*, the sister ship to the massive Japanese battleship *Yamato*.

Allen realized the *Octopus* would not suffice as a platform for finding ships from the war. It was a luxury yacht, not a research vessel, though it doubled as both. Consequently, Vulcan purchased a ship originally called the *Seven Petrel*, a 250-foot-long dynamically positioned ship ideally suited for deep ocean exploration. The ship was renamed the RV *Petrel* and was soon outfitted with the most advanced underwater technology in existence, including a deep ocean Kongsberg multi-beam sonar (which can scan the sea bottom in great depths), a 6,000-meter-capable Remus AUV, and a Norwegian-built Argus ROV specially outfitted for Allen's requirements. Apparently, Allen also believed in keeping his people happy because the *Petrel* was also fitted with the most expensive coffee makers and deck stereo systems I had ever seen. His crew could rock out on deck while they worked and did maintenance. It was awesome.

It was a significant investment, and over the years the *Petrel* racked up one discovery after another, much to the chagrin of Dr. Robert Ballard (of RMS *Titanic* fame) as the *Petrel* and her skilled team knocked every one of his beloved targets off his list. In addition, the *Petrel* had the ability to stream four real-time HD-quality video channels via satellite, technology that would be demonstrated during the live survey of the USS *Indianapolis*. As a result, Allen could sit in his office in Seattle (or probably also his home) and direct survey operations in real time. No one else in the world had such a capability.

Allen was serious about finding shipwrecks, and find them he did, as the *Petrel* found and surveyed every historically significant wreck lost in the Pacific theater, including the *Indianapolis*, carriers *Lexington* and *Hornet*, the *Ward* (which fired the first shot at Pearl Harbor), all the Japanese carriers from the battle of Midway, and the USS *Juneau* of Sullivan brothers fame. These were amazing accomplishments.

But none of this had happened yet when I first heard rumors that Allen was going after the *Indianapolis*. I was spending my free time going over old stacks of Navy documents to make sure my analysis was correct. Fortunately, the underwater community was a small one, and I was referred to Robert Kraft, Allen's director of subsea operations, by Gary Kozak, a well-known expert in analyzing side-scan sonar records. Gary and I became good friends on the *Petrel*, and he is still doing the work today.

Looking back on the expedition seven years later, I really admire what Paul Allen accomplished. He could have sat on his billions of dollars and done nothing. Instead, he took a substantial financial risk to find the *Indianapolis*. Granted, to a billionaire, maybe $20 or $30 million to buy the ship and the vehicles, hire the people, and so forth was not a lot of money. But Allen really was the only person who could have done such a thing. No one else was willing to risk so much money to find the ship, but he was, and did. Also, the guys working for Vulcan on the *Petrel* were as smart and dedicated as any people I had worked with in forty years. But what really impressed me was that when Allen publicly discussed his expeditions, he always said *we*, not *I*. That meant he understood that it wasn't just him doing the expeditions, but his team. In my opinion, Paul Allen deserves all the credit for everything the *Petrel* found.

I think it was in late 2016 that Vulcan flew me down to Underwater Intervention in New Orleans to meet and discuss my research. The Vulcan group were very secretive and told no one about their operation or plans. I was fine with that, but it made it difficult for me to determine what their plan was so I could adapt my analysis. At that time, the *Petrel* was in Norway doing sea trials with Vulcan's new Remus 6000 AUV, still working out various bugs and finalizing procedures. I had seen that vehicle in the final stages of construction at Hydroid Inc. in Massachusetts earlier that year. Even though I was poised to purchase a $6 million system for the Navy, all they did was smile and look confused when I mentioned Robert Kraft's name.

During our lunch, I showed them my plots and told them everything I knew about the area and the circumstances of the sinking. There was no point in hiding anything I knew because I needed to convince Vulcan that I could help them in their search. Of course I had to sign a nondisclosure agreement and that was fine. It was nobody's business what I or they were working on.

After many discussions with Tom Grieshaber, one of the Seattle-based managers, I was hired as a consultant to the *Indianapolis* expedition, along with Gary Kozak. One sticking point in our discussions was the ownership of the intellectual property of my existing research. Vulcan couldn't figure out a way to pay me for previous work, so I agreed to give them full access to this data and in return I retained the IP rights.

As the *Petrel* steamed out of the Mediterranean Sea and through the Suez Canal, I watched the team's progress on shiptracker.com. Allen was a bold visionary, and his first target using his new toys was the most famous shipwreck left undiscovered. It was a gutsy move because the *Indy* was hands down going to be the most difficult target to find as well due to the lack of a decent sinking location and the rugged terrain in the area. But the *Petrel*'s team were very, very good at what they did and were determined to be successful. Kraft and his people had nearly an unlimited budget for the search, which was a good thing because it was going to be expensive. In fact, the head of the Naval History and Heritage Command (NHHC) was reportedly stunned when Kraft told them that Allen was going after the *Indianapolis*. From what Kraft told me, the conversation went something like this:

Kraft: "Paul has told us that he wants to find the USS *Indianapolis*."
NHHC: "How long do you plan to look for it?"
Kraft: "As long as it takes, or until I am directed to do otherwise."
NHHC: Silence.

That silence was because it was well-known that the NHHC was in bed with Robert Ballard, who also wanted to find the *Indianapolis*. But Ballard had nowhere near the budget that Paul Allen did. Allen was determined to find the ship and, if Vulcan had not discovered the *Indianapolis* when they did, it would probably still be unfound because no one else could afford to spend "as long as it takes" to find it. How do I know that Ballard was planning to look for the *Indy*? Because I was asked by the Navy to do a cost estimate for just that using government-owned equipment. They looked puzzled when I told them it was a waste of time.

By the time I departed Dulles Airport for Guam on May 15, 2017, I was still in bad shape emotionally. Although I had been able to do a lot of analysis on the *Indy*, I was not well. I had been on the antidepressants for over a month, but they had not kicked in with the initial 5 mg dose. Consequently, I was just not my usual self. Of course, even my usual self wasn't exactly a people person. However, I was hopeful that the trip might snap me out of my depression. It really didn't, but I managed to struggle through. Even though I was part of Vulcan's expedition and was being well paid (and I desperately needed the money after the divorce judgment), I was very unhappy and there was little I could do about it.

I didn't know what to expect when I arrived in Guam. On the flight with me were an engineer from Hydroid and a few other Vulcan contractors. I didn't know any of them, so I felt out of place. But before long, we got our ride to the ship, which was docked at the same pier as the MV *June T* had been docked at almost twenty years earlier when I did my search. But the port had changed and had much tighter security now. Fortunately, I brought along my US Navy contractor access and transportation worker identification cards and was able to escort people in and out of security.

Before long, I was sleeping in my rack on an unknown ship waiting for it all to begin. I met everyone else during our fire and lifeboat drill, among them Robert Kraft. I didn't know much about him, but he was definitely in charge of things. I also again saw Gary Kozak as well as Paul Mayer, Vulcan's lead researcher.

One thing I learned from the outset was that if you were a contractor, like Gary and me, you weren't really part of the team. We were from the technical standpoint, but we dined at tables separate from the Vulcan guys. It wasn't a real rule, but apparently protocol. But when it came time for happy hour on the back deck with our beer, we all sat and joked together.

There were only a few full-time Vulcan employees on the project: Robert Kraft, Paul Mayer, and Wayne, their networking engineer. Others were on long-term contracts, such as Pat, a small-game hunter from the UK who owned the most expensive rifle scopes I had ever seen; Rudy, their fabricator from the Bay Area in California who could build anything and was a real craftsman; and Eric, formerly of Bluefin Robotics. Eric was kind of standoffish at first, but we became friends during the search. We also had a film producer on board, Rob Lyall of The Biscuit Factory from Falls Church, Virginia. He and I would spend much time doing interviews and talking about the *Indy*.

The *Petrel* was not a new ship, but Paul Allen had spared no expense in outfitting it with the latest in underwater technology and creature comforts. The food was superb, and I gained 10 pounds in six weeks, which is typical of when I go to sea on a long job. There was beer and wine to drink (hard alcohol was not allowed), an exercise room, and a streaming video library that allowed us to watch movies from our cabins.

Vulcan created an operations center second to none. The Remus AUV and Argus ROV were controlled from one room on the main deck that contained racks and racks of assorted equipment, as well as a slew of large-screen video monitors where you could monitor the deck cameras and review sonar or navigational data. I was surprised by their use of Google Earth as a primary navigational plotting device. Certainly, they used other mappings applications as well, but Google Earth was an excellent tool to do general plotting and allowed everyone to see what the plan was.

One deck below was the "offline room," where everyone had their desks and personal computers. Several monitors were set up, which allowed everyone down there to plan operations and monitor the progress of either vehicle when it was at depth. Down the hall was Rob's office, as well as a large conference room where we held our meetings and discussed the search plan. Operational planning sessions began soon after I got on board. During these, we all expressed our thoughts on where to search and why.

By the time we left Guam on Sunday, May 21, 2017, we had already had several planning meetings, mostly to review the attack sequence. While we were examining all the available data, the *Petrel* headed 15 miles off the coast to calibrate their HiPAP (high precision acoustic positioning) navigation system, crucial to tracking both the Remus AUV and Argus ROV. After only a day offshore, we returned to give a tour to several US Navy officers based on the island. Before steaming to the estimated sinking location, Vulcan's team conditioned their ROV cable, in all places the Mariana Trench. Conditioning an ROV umbilical, especially one made from steel wire, is important to make sure all the turns have been removed and the cable is wrapped back on the reel under tension. Given the 11,000 meters of water depth, we certainly didn't have to worry about dragging the vehicle's termination on the bottom. During this time, we also ran several AUV missions with the Remus to check the sub's pitch and yaw stability.

The three people who did most of the analysis on the search area were me, Paul Mayer, and Gary Kozak. To a lesser extent the NHHC were involved, but none of us liked their analysis because it contained several errors. Myself, I favored a strictly navigational study using the *LST-779* sighting as a starting point. We had good data on the course and speed of the ship, and it was an easy mathematical solution predicting where the *Indy* probably was at a given point in time. I felt that the survivor locations were useful only in determining the farthest west boundary of the search area and thought it was too difficult to model the leeway of the survivor drift over five days.

Mayer, on the other hand, took a different approach. He liked the survivor data and spent a lot of time plotting the locations of where survivors were found. I had done that in 2000, but many of the debris locations were from dead reckoning fixes and thus dubious. The only decent fix Mayer had for the survivors was the Chuck Gwinn 1130 hours LORAN-A fix. But even that could be off by several miles because Gwinn was at the edge of his reception when he recorded the location.

Kozak was interested in the *I-58* position because the survivor drift line led directly back to Hashimoto's estimate. It was not the full basis of his thinking, but he raised the fact during several of our meetings. I thought it was interesting, but I couldn't see how the ship could be that far off track. I felt that the ship was likely within 5 to 10 miles of the Peddie route, somewhere west of where I had searched in 2000 and east of where the survivors were found. Of course, that still covered a lot of real estate, which was why it was so important to be able to spend sufficient time searching a large area.

We didn't start steaming to the attack location until May 31, after receiving spare parts for the ROV's electric winch, which was already having problems. These issues would come back to haunt us later.

As we steamed at 10 knots along the *Indy*'s 1945 path, one surprising discovery was a massive 7-kilometer-deep depression in the bottom. This anomaly was revealed by the *Petrel*'s multi-beam sonar, and it almost looked like an ancient meteorite crater.

We finally arrived at the search area on June 2 and began using the ship's multi-beam sonar to collect bathymetric data. This was critical for safe AUV operations because it would determine the orientation of the vehicle's search lines. The bottom in the area alternated between 4.5 and 5.2 kilometers of water depth, with sediment-filled basins and hard rock outcroppings. If you can imagine it, it was almost like searching through the Grand Canyon, except this canyon was miles underwater, freezing cold, and in pitch-black darkness. When the *Indianapolis* sank, they lost the game of Battleship with the *I-58* and paid for it with hundreds of lives. We were also going to be playing Battleship, but our version would be picking the right block of ocean floor that contained the ship.

Initial Remus AUV operations were centered in an area composed of twelve individual search boxes, most of them about 10 kilometers square; the bulk of them were south of the Peddie route, with four or so to the north. The Remus left the surface on the first dive at 1900 hours on June 8, 2017, to start its long journey to the bottom. Vulcan had developed a slick descent procedure using a 60-kilogram descent weight in the vehicle's nose. Once the AUV was released from the crane, it sank like a stone until the weight dropped at 500 meters off the bottom, allowing the vehicle to drive the rest of the way down. It would be bad if the weight didn't release, which happened during one later dive.

Once the Remus dropped the descent weight and drove itself toward the bottom, it wasn't long before there were problems. An AUV senses its altitude above the bottom and speed in any direction using a Doppler velocity log. The device is essentially a sonar with multiple beams that reflect off the seafloor, and this data is used to calculate speed and distance. The problem we were having was that the terrain was so steep that the Remus could not stay close enough to maintain bottom lock. This was because the vehicle's control system could not pitch the vehicle down fast enough to stay sufficiently close to the sediment as it plunged down the hills. It wasn't a problem going up hills, only down, because the bottom rapidly fell away from the vehicle as it drove forward. Once the vehicle lost bottom lock, it would abort the mission, requiring a long ascent to the surface. That took a lot of time, and as in anything, time is money. The onboard engineer from Hydroid eventually solved the problem by reprogramming the vehicle to dive at steeper angles down the terrain. What we found was that the geology was horrific: changes in depth from 3,700 to more than 5,200 meters in one area, and these areas were totally unexplored. That's an elevation change equal to the height of almost nine Washington Monuments.

On June 12, we decided to shift our focus to another area that was slightly north of the Peddie route. The vehicle made it to the bottom without incident, and after downloading the sonar data on the first dive, we discovered a ship-like target 120 meters long and 20 meters wide surrounded by a small debris field. Vulcan quickly reprogrammed the Remus to use a range scale of 800 meters per side, but it aborted before reaching the bottom.

After recycling things (i.e., doing everything necessary to prepare the AUV to dive again), we dumped the AUV and commanded it to make several high-resolution passes on the mysterious contact. Did we think it was the *Indianapolis*? Probably not. It was too small. But it could be the SS *Nanman Maru*, a Japanese troop transport torpedoed on October 27, 1943, by the USS *Flying Fish* (SS-229) during her seventh war patrol. Incredibly, the ship was only 4.5 miles from the sinking location as reported by the Imperial Japanese Navy; turns out the Japanese were pretty good navigators.

The ship was never positively identified using the Argus ROV, but I was confident in my identification because the target's characteristics matched those of the SS *Nanman Maru* very closely.

The Remus dive on June 20 started like any other, with the yellow torpedo-shaped vehicle plunging into the abyss like a chunk of lead. But this time, the heavy descent weight didn't release. Instead, the Remus smacked into the sea-floor like a bag of rocks, anchored on the bottom by the weight. Once Kraft's team figured out what had happened, they sent an acoustic command to the vehicle to drop the emergency ascent weight. In a cloud of sediment in the ink-black darkness, the weight indeed dropped; free of the extra mass, the $6 million Remus began a long and torturous ascent to the surface.

Unfortunately, at a depth of about 1,000 meters, the ascent stopped. The vehicle had reached a layer of water in which it was neutrally buoyant, caus-ing it to drift at mid-water like a jellyfish. By the time all of this happened, it was dark, and Vulcan was faced with a real dilemma: the Remus was not on the bottom, but it wasn't on the surface either. There were no iridium mes-sages nor strobe light to mark her location—just the steady pings from the Remus's HiPAP acoustic beacon.

After a short discussion with the team, Rob ordered the Argus ROV into the water, fitted with a sickle-shaped knife in her starboard manipulator. This was not going to be easy due to the movement of the various players in the dive. First, there was the AUV, slowly sinking and rising in a layer of less-dense saltwater. The vehicle was not at a constant depth. Instead, it would rise, lose buoyancy, sink until it hit the bottom of the layer, then rise again. It was like a very expensive yo-yo drifting in the sea. Then there was the *Petrel*, with the Argus ROV maneuvering below at the end of her armored umbil-ical. As a result, there were three entities involved: the Remus, drifting hap-hazardly in the darkness, the *Petrel* on the surface, and the ROV, being driven under the ship. All three would have to be in exactly the right orientation to pull off the rescue.

After about an hour of searching, the Argus detected the Remus on her BlueView scanning sonar and the two triangular icons on the WinFrog naviga-tion screen merged. Out of the gloom, the AUV revealed herself as the Argus was slowly and carefully maneuvered until the pilot snagged the drop weight line with the razor-sharp knife. With an abrupt action, the polypropylene line was sliced and the AUV soared to the surface. While the rest of the recovery was routine, it had been an impressive demonstration of operational skill. In the end, the mishap was caused by an abnormal orientation of the drop weight chain in the release mechanism. It would never happen again.

Unfortunately, it would not be long before several technical issues forced a temporary halt to the search. First, the two massive electric motors that rotated the ROV's cable reel began to overheat. In addition, the drive mechanisms between the motors and chain were continually becoming misaligned, causing the constant velocity joints to melt, then fail. But what was even worse was that several fasteners on the overboarding sheave had become dislodged and caught underneath the armored steel umbilical. This damaged the outer wires of the umbilical, and we spent several hours taping up the exterior of the cable with Kevlar tape. But there was no fixing it. The cable would have to be replaced.

Finally, the Remus AUV began to experience problems with its propulsion system (i.e., the tail cone), and the Vulcan team eventually ran out of spare parts. With no more spares, the *Petrel* came out of dynamic positioning and charted a course for Palau. My role in the search was now over because Vulcan had a limited amount of funds to pay me and there was no more to be had. Honestly, I would have stayed out there for free; that was how bad I wanted to be there when they found the ship. I knew it was only a matter of time because the expedition was not over, just suspended for technical reasons. In a little more than a month, we had searched more than 470 square miles in water depths down to 5,200 meters and found one ship; it just didn't turn out to be the USS *Indianapolis*. But that would soon change.

On the morning of August 19, I received a cryptic phone call from Maria Bullard of the Second Watch Organization, telling me that she had heard a rumor that the ship had been found. I had heard nothing from my friends at Vulcan. But later that night, I received an email from a member of the Vulcan team whom I had met during our earlier search, with the subject line "FOUND IT! FOUND IT!" Where the *Indianapolis* was found is a closely guarded secret. However, Paul Allen eventually put out an official press release, and it wasn't long before the internet was flooded with various images from the deep. There was no question about it. The ship had finally been discovered.

Since that day, two PBS documentaries have aired showing the ship in great detail. The last one was a live broadcast from the wreck site—that alone an amazing feat to pull off. I was featured in both of them, in the first as an expedition consultant and the second as a historian. However, the important thing is that the *Indianapolis* was discovered and the location of the abyssal gravesite of several hundred men was marked in history. It was an amazing accomplishment for all who worked on the project.

16

2023: Race to the Deep Ocean

While the United States rests upon her sixty-plus-year-old laurels, China has mounted an aggressive and long-term assault on the deep ocean, or the region known as the hadal zone, areas deeper than 6,000 meters (19,685 feet).

Our country's efforts to explore these depths effectively ended in 1960 at the conclusion of Project Nekton. That was when the US Navy's *Trieste I* bathyscaphe reached the deepest known part of the ocean, the Challenger Deep, with Lt. Don Walsh and Swiss explorer Jacques Piccard inside a tiny steel sphere.

Some of China's current efforts are being led by Prof. Cui Weicheng of Shanghai Ocean University and, more recently, the Rainbowfish Ocean Technology Co. Their plan was to deploy 11-kilometer (6.8-mile)-capable submersibles, manned and unmanned, in 2020 to explore the deepest sections of the Mariana Trench. According to the *South China Morning Post* in 2019, "Beijing has listed deep-sea scientific exploration as one of the key projects in their five-year plan to 2020." They have accomplished this objective.

These programs involve the development of unmanned vehicles, benthic landers, gliders, and manned submersibles, as well as the infrastructure needed to maintain this technology. In addition, several research vessels have been constructed to support field operations, most of them purpose-built for specific systems.

How have the Chinese created such capabilities? Their projects were initially made possible through the purchase of non-ITAR (International Traffic in Arms Regulations)-controlled underwater sonars and vehicles from British, Norwegian, and Canadian companies. (ITAR governs the export of US defense and military products.) As has been the case before, Chinese engineers used this technology as an educational tool so that they could copy and improve on existing designs.

The next phase was the outright purchase of large engineering firms developing commercial off-the-shelf remotely operated vehicles (ROVs). To this end, Chinese companies have acquired UK-based Soil Machine Dynamics Ltd. and the Deepflight Submarine Co. The acquisition of SMD by China's Zhuzhou CSR Times Electric Co. is especially troubling because SMD was a major supplier of underwater technology to academic, commercial, and potentially military operators in the United States.

In the last few years, the Chinese have concentrated on developing their own underwater technology through government-funded research and development and what information they can glean from numerous technology conferences focused on the subsea market. Their capabilities developed slowly as they designed and built increasingly more complex and depth-capable systems, such as the Sea Pole class of bathyscaphes; the *Haidou I* unmanned vehicle, which reached 10,767 meters (35,324 feet) in 2016; and the DSV *Jiaolong*, which can carry occupants to depths as great as 7 kilometers (4.3 miles) underwater. The stainless steel personnel sphere for the Rainbowfish vehicle was forged in Finland at Tevo Lokomo and is large enough to carry three people. They planned to begin testing this vehicle in 2020 from their 4,800-ton research ship *Zhang Jian*.

Surprisingly, the People's Republic of China has also enlisted the technical assistance of several American and European experts in underwater technology to further their program, such as Dr. Don Walsh (former *Trieste* bathyscaphe pilot), Dr. Sylvia Earle (Ocean Everest program), director James Cameron (developer of the DSV *Deepsea Challenger*), the Deep Submergence Laboratory at the Woods Hole Oceanographic Institution,

Dr. Anatoly Sagalevich (director of Russia's Mir program), and Dr. Alan Jamieson (senior lecturer at Newcastle University and chief scientist on the Five Deeps Expedition).

A critical question is whether the Chinese will continue to buy their deep ocean capabilities. If they were to acquire one specific Norwegian company, they would seriously hinder the ability of the US to work in the deep ocean. That company is Kongsberg Maritime (KM), based in Horten, Norway. They manufacture a large fleet of autonomous underwater vehicles (AUVs) that are used by several European navies, as well as our own US Navy.

At this time, the capabilities of the United States' unmanned submersibles (the CURV 21 and *Deep Discoverer* ROVs) are limited to 6,000 meters (19,685 feet), while the manned DSV *Alvin* can reach 4,500 meters (14,763 feet). Woods Hole's *Nereus* hybrid vehicle was lost in 2014 when it imploded at a depth of 9,900 meters (32,480 feet) in the Kermadec Trench.

There has been more exploration of the hadal zone by private American citizens since Project Nekton ended. James Cameron used his *Deepsea Challenger* manned submersible to dive to almost 11,000 meters (36,089 feet) in 2012. Unfortunately, his vehicle made only one trip and has since been donated to the Woods Hole Oceanographic Institution; it is no longer operational. Wall Street trader Victor Vescovo amassed an impressive record of multiple dives to over 10,000 meters (32,800 feet) with his DSV *Limiting Factor*; he went to the bottom of the Mariana Trench an unprecedented five times. After that, he explored all of the Earth's deep trenches using his support ship, the DSSV *Pressure Drop*. However, his Five Deeps Expedition is now complete, and he has sold both the submersible and research ship to recoup his investment.

There are several possibilities for why China is embarking on such a costly undertaking:

Scientific: To study samples of hadal zone sea life (primarily microorganisms) in support of their pharmaceutical industry.

Commercial: To sustain their claim in the Clarion-Clipperton Zone for the mining of polymetallic nodules (manganese). The International Seabed Authority estimates that the total amount of nodules in this area exceeds 21 billion tons (Bt), consisting of 5.6 Bt of manganese, 0.27 Bt of nickel, 0.23 Bt of copper, and 0.05 Bt of cobalt.

Strategic: To further develop technology that can weaponize the deep ocean
on both tactical and strategic levels—for example, potential systems like
the US Department of Defense's Defense Advanced Research Project
Agency's proposed upward falling payloads. These payloads consist of
defensive and/or offensive weapons as well as unmanned aerial vehicles
for interdiction and surveillance. UFP systems deployed beyond 6 kilometers (3.7 miles) will be out of our reach.

During my last year working at Phoenix International Holdings, I was
tasked by the US Navy's Supervisor of Salvage and Diving (SUPSALV) to
develop a plan to extend their search and recovery capabilities to 11 kilometers. This came about following a directive from the chief of naval operations, OPNAVINST 4740.2H, published January 7, 2021, in which it states
that we are to "maintain and operate deep ocean search and recovery assets
with global full ocean depth capability down to 11,000 meters (36,000 feet)
sea water."

I came up with a variety of options, including single-use disposable inspection vehicles and having an AUV developed by a subcontractor. I even offered up my own conceptual design that used many off-the-shelf components.
Surprisingly, there are 11-kilometer-rated subsystems available. Not many, but
they exist. The design I created would have used several lithium-ion battery
modules and a well-tested side-scan sonar and imaging system, all of it cost-
ing $10 to $12 million.

Unfortunately, nothing came of it and SUPSALV considered my work as
completing their assigned task. The likely reason nothing came of it is that
they didn't have the money to go down into the hadal zone. We were having
enough trouble finding the funds to purchase an autonomous search vehicle
to replace our aging intermediate towed search systems.

It is not that the United States doesn't have a deepwater presence. We do.
But our ability to function in the deep ocean stops at 6,000 meters (19,685
feet), while China has made major strides and broken through this barrier
into the hadal zone. Is it expensive? Yes. The question is whether we can afford *not* to go there, because whoever controls the deep ocean also controls
the waters above.

17

2024: Reflections

M y brother, Chris, and I were lucky in that we had good parents. They weren't perfect, but they did a good job of raising us and instilling within us a sense of purpose and values.

Our mother was a wonderful woman who always wanted the best for Chris and me. She was a faithful Army wife who took care of us and the household while Dad was with the 4077th MASH in Korea, mapping Alaska, and serving his two tours in Vietnam. I was always a handful to control, but I loved her for trying.

In the early 1980s, I lived with my mother in Vienna, Virginia, while I studied electronics engineering technology at Capitol Institute of Technology (now Capitol Technology University) in Maryland after leaving Ocean Search. During a trip to California to visit her sisters, she somehow fell off a chair and broke her arm. When I picked her up at the airport, I commented on how tired she looked. "What did they do to you out there?" I asked her. She had her arm in a sling and just appeared exhausted.

Over time, we noticed a growing weakness in her mouth and vocal cords and that she was having problems speaking. Other portions of her body were weakening as well, such as her hands and legs. Clearly, something was wrong. After a battery of tests, she was diagnosed with bulbar palsy, a motor neuron disease. As we would discover later, what she really had was amyotrophic lateral sclerosis (ALS). Over time, her condition deteriorated to the point where the loss of control of the muscles in her throat made it almost impossible for her to eat and she had to be fed through her stomach. It became obvious that she needed to return to California, where she could be cared for by her family. I always held out hope that she would get better, but these hopes were dashed one day by my Uncle Charlie when he said to me, "You know, she's not going to be around much longer." I had a very difficult time accepting that.

Mom died in April of 1985 after suffering a respiratory arrest. I'm grateful that she was not alone when it happened but was watching television with her brother Johnny in the family home at 591 Capell Street in Oakland, California. I took her death very hard.

My father was a more complex person than Mom. He never had the advantage of being raised in a stable household, as did my mother. Despite the difficulty of his childhood, he managed to put himself through college, become a professional US Army officer and aviator, survive three wars, and raise a family. He never liked to talk about his childhood. I think it brought back too many painful memories.

When I was working for Ocean Systems in Houston, Texas, I, too, learned to fly. A couple of times, I flew from Andrau Airpark up to Huntsville, Alabama, where my dad was living with his new wife, Dorothy, to meet them for lunch. On the second trip, I rented a Piper Cherokee 150 and landed at a small grass runway near his house. Dad and I went flying together, and at one point, I let him take the controls. You can tell a lot about a pilot just by the way they caress and manipulate the controls. It was obvious that my father still had the gift, even though he admitted that flying did not come easy for him.

What's funny about Dad is that the more I learned about him, the more I started to know myself because we were one and the same in many ways. In short, I was Version 2.0 of Elswick "Newt" Newport, for better or for worse. We had many of the same flaws and strengths.

Dad lived to ninety-six years of age and died a month to the day after his birthday. I was at his bedside shortly before he drew his last breath and confessed my sins to him, apologizing for running away from home and telling him that I loved him and that he was a good father. I hope he heard me. Later, I found microfiche copies of his Army records, which contained a wealth of information about him, including his efficiency reports. They showed that he was an exemplary officer who would have been destined for promotion to general staff officer except for one thing: his slight stutter. For this one defect, the Army discriminated against him. Even so, he was a courageous and skilled Army aviator, as demonstrated by his award of the "Air Medal with Ninth Oak Leaf Cluster with 'V' device" as part of the Army Concept Team in Vietnam (ACTIV), 1965–1966:

> Lieutenant Colonel Newport distinguished himself by exceptionally valorous actions on 19 February 1966 while serving as the pilot of an armed helicopter on a reconnaissance mission near Duc Hoa, Republic of Vietnam. . . . The mission was to determine the effectiveness of an experimental 2.75 inch anti-personnel rocket. . . . Newport flew his aircraft at an extremely low altitude to reconnoiter the effectiveness of the rocket warhead. This low level flight resulted in the capture of a Viet Cong prisoner. After delivering the prisoner to a rear area, he returned to the operational area. . . . Reconnoitering the area at a very low altitude and under intense ground fire, Lieutenant Colonel Newport searched for additional cratering effects of the warheads and for more Viet Cong casualties. Although his aircraft received several hits, he continued his search, landed again, and captured another prisoner. As a direct result of his exceptional flying ability, courage, and determination, two Viet Cong prisoners, a sub-machine gun, a carbine, grenades, documents, and numerous other intelligence valued items were captured.

What the above narrative does not mention is that during one of his landings in his UH-1 Huey, Dad was ambushed by three Viet Cong who sprang up out of the elephant grass. My father did not carry the normal military service pistol, but a powerful Smith & Wesson .357 caliber magnum Model 27 revolver. Unfortunately for these VC, Dad was excellent with a pistol; he shot all three of them and survived.

While sitting around the dinner table one night in Colorado Springs, I asked Dad if he ever used this pistol in combat. His reply was brief: "Only once, and it worked."

Despite all those years of flying, being shot at, and shooting at others, Dad never received an injury, other than a minor shrapnel wound following a mortar attack in Korea, and of course being dunked into San Francisco Bay when he ditched his Army Navion. Clearly someone, or something, was looking out for him. I think about both him and Mom every day.

Thinking of the *Liberty Bell 7* project over twenty years later is a poignant experience. I say this because, sadly, about half of our team are no longer living, all younger than me. I still haven't figured out why I'm still around, except possibly for the genes I inherited from my father. Those who were taken from us far too soon were Ron Schmidt, Richard Daley, Steve Wright, and Dave Warford. They were all good men and experts in their chosen field. Both Jim Lewis, the pilot of *Hunt Club 1*, and Guenter Wendt, the famous NASA pad leader, have also left this world.

I think part of this has to do with the physical and mental stress related to the profession. All of the international travel, time changes, shipboard living conditions, bad food, back-breaking work, and around-the-clock operations take a toll on a person. Drug and alcohol abuse are no stranger to this profession, mostly to relieve job-related stress.

Weeks and months spent away from home do not result in a happy family life, and most marriages end in divorce. Absence does not make the heart grow fonder. The hypothetical "Jodys" are the people who service the wives while their husbands are at sea. They are not hypothetical because I had one of my own, whose existence was established by the empty liquor bottles I observed in the trash bin after coming home from Navy work at sea. Women have wandering eyes as well. Children? They typically get to know their fathers via satellite phone calls.

Given all of this, one must ask the question, *Why do people do it?*

The subsea professions draw an eclectic group of people. In some cases, they want to see the world, but that gets old after a few dozen international flights. Sometimes, they are running away from an unhappy life, such as a bad marriage. These usually take care of themselves, and more than once people have returned home to find a house or apartment empty of furniture and a bank account closed. There is also the adventure of it all and having the opportunity to join a very select group of people who are given the chance to see things many people only dream of. And there is the opportunity to work in and contribute to a highly technical field and hone their skills. If someone can troubleshoot and repair an underwater vehicle on deck, they can generally fix anything.

Work at sea is difficult, and in some ways we are not unlike professional athletes; we have a limited window within which we can do the work due to the physical demands of the job. Of course, none of us are paid multimillion-dollar salaries, and we don't have the health benefits of such a person. Myself, I stayed in the game far too long and paid the price with ongoing health issues that will be with me for the rest of my life. I didn't want to do this but was forced into it by an excessive alimony requirement during my divorce. It was either keep working at sea or go broke. I still suffer from serious lower back pain from the twelve-hour-a-day shifts and physical work. I simply wore out my back after more than forty years at sea. My hearing is also pretty much shot after so many years on noisy salvage ships. In fact, I suffered an almost total loss of hearing in 2022; it is called sudden deafness, or sudden sensorineural ("inner ear") hearing loss (SSHL). It happened over a period of about thirty minutes. I was driving home from work and thought that something was wrong with my car because it sounded different. It wasn't the car; it was my hearing. Fortunately, after a two-week steroid treatment, most (but not all) of my hearing returned, though I am still afflicted with tinnitus, a constant ringing in the ears. Most of my hearing loss is at the higher frequencies, which is common with age anyway. I still suffer from insomnia and at times have had nightmares so severe that I have rolled out of bed in my sleep. Many of our band of brothers are veterans of a range of militaries, such as the Marines, Navy, and Army. Those of us who are not, like me, receive no special benefits or considerations for our decades of service to our country as employees of a US Navy contractor. On several occasions, I was told that I was "in the Navy." My reply has always been the same: "Where's my pension?"

I am not suggesting that I have had a hard life. I think I have had an easy life. I was raised in a stable family environment by two loving parents, never went hungry, and always had shelter. Any problems I have had in life were generally self-inflicted.

Anyone who says they have no regrets in life is either lying to themselves or has not lived much of a life. I have so many regrets for the decisions I made that they could fill a whole other book.

I regret dropping out of college because I would have been an excellent engineer. But at the time, I simply didn't know what I wanted to do. Granted, I have done engineering work during my career, but I didn't have the diploma. In fact, I have always felt that I knew more about certain engineering aspects of underwater vehicles than most degreed engineers. Why? Because I know what works and what doesn't.

I never should have run away from home. It was the coward's way out. I should have told my parents that I was being bullied at school and let them handle it. I should not have subjected my father to the embarrassment of having to take emergency leave from the Pentagon to drag me back home from California. What was I thinking? My plan was solid, but long term, it was a fool's errand. What was I going to do in California? How would I live? I never figured out that part.

I curse myself for not being there for my mother during the last months of her life. But her sister, Mildred, put me in a position where I had to leave our family home on Capell Street in Oakland because she viewed me as a threat to her authority. That was not my intent; I just wanted to be with my mother because she was sick. I should have stood up to her and been there for Mom when she died.

My brother, Chris, joined the Army as an officer, became a paratrooper, and ended up in the Berlin Brigade in Germany. After about five years, he became disillusioned with the service and resigned his commission. I believe Chris lost his way in life over the years after suffering some serious health issues. For a while, Dad and I weren't certain he would survive. But he did, and he now lives in California with his wife, Sylvia, and their daughter, Diana. Chris is an excellent writer, and I hope he finds his path.

I, too, have a daughter, one created in 1992 with Eija Lippo, a Nordic beauty who was then working as a White House correspondent for the Finnish section of the BBC. She is now semiretired from journalism after working for the local Porvoo newspaper for many years. She was in attendance when my father was interred at Arlington National Cemetery, with full military honors, in April of 2022.

My daughter, Milana Newport, has grown up to be a very intelligent, logical, and attractive young woman and is currently studying chemistry at the University of Helsinki. We communicate one way or another on a near daily basis. Milana lives in Espoo with her fiancé, Michael, a handsome young man who I am certain keeps her pointed in the right direction.

I feel guilty about not having a relationship with Milana during the first nineteen years of her life. At the time, the internet was in its infancy and email was nonexistent. I should have tried to communicate with her but didn't. Later, when Milana was older, I was married, and acknowledging her as my daughter would have caused more problems in an already difficult relationship. But eventually I came around, mostly due to my father's efforts to bring us together. My father's words were to the effect that if it's the last thing he did, he would bring Milana and me together. Mission accomplished, Dad.

Over the years, there have been trips to Helsinki, Porvoo, and Espoo, Finland, short excursions added on to scheduled Navy travel. And both Eija and Milana have been over here. We once met in Colorado Springs to visit my father.

I also regret not enlisting in the US Army when I was on the verge of being drafted in the early 1970s. The discipline would have been good for me, and I would have made a competent helicopter pilot in the Army's Warrant Officer program. At the time, my father assured me that he could get me into any program I wanted. I should have taken him up on his offer and flown helicopters. Maybe I could have saved some lives in Vietnam flying a helicopter ambulance like Dad. But we'll never know because at the time, I was tired of living on Army bases.

To be honest, I didn't really like working on ships. I once calculated that I have spent seven to ten years of my life at sea. There's nothing romantic about it: the work is long and hard, hazards abound, the food is not that good, and the accommodations are noisy and hot. Like it or not, we were the last responders. We were the guys who headed out to pick up the pieces after everyone else had gone home.

So then, the question must be asked. Why did I do it? The answer is fairly simple. I did it because that was where the action was with respect to underwater vehicles. And besides, where else could one man have had such adventures in the abyss?

Maybe this is the only decision I've made during my life that I do not regret.

Acknowledgments

First, I thank my parents for their love and allowing me to pursue my dreams. I miss them both every day.

Thank you to Justin Race, Chris Brannan, Bryan Shaffer, Kelley Kimm, Katherine Purple, Andrea Gapsch, and the rest of the staff of Purdue University Press for their support during the long development of this book.

Sincere thanks to former Phoenix International Holdings coworkers Patrick Keenan, Matt Long, Justin Hall, John Finke, Mark Bender, and Josh Konicki for helping to refresh my memory as I wrote about certain operations.

Special thanks to George Leopold, author of *Calculated Risk: The Supersonic Life and Times of Gus Grissom*, my daughter, Milana, and her mother, Eija Lippo, for reviewing certain sections of the manuscript and providing their insight.

And finally, I thank all of my friends and former coworkers who preordered this book. It all helps in the end.

Curt Newport
Operational Experience
1977–2020
(partial listing)

Note: Operations denoted in **boldface** are discussed in this volume.

Abbreviations: *ADS*, advanced diving system; *ADV*, Australian Defence vessel; *ALFS*, airborne low frequency sonar; *AUV*, autonomous underwater vehicle; *CANTAT*, Canadian transatlantic telecommunications; *CCGS*, Canadian Coast Guard ship; *CS*, cable ship; *CURV*, cable-controlled undersea recovery vehicle; *DB*, derrick barge; *DOSS*, deep ocean search system; *DP*, dynamic positioning; *fsw*, feet of sea water; *MALD*, miniature air-launched decoy; *MAT*, Mediterranean–Atlantic; *MV*, motor vessel; *PACAF*, Pacific Air Forces; *PLBO*, post-lay burial operation; *PM*, project manager; *ROCAF*, Republic of China Armed Forces; *ROV*, remotely operated vehicle; *RV*, research vessel; *SAT FADS*, saturation fly-away diving system; *SCARAB*, submersible craft assisting repair and burial; *SCORPIO*, submersible craft for ocean repair, positioning, inspection, and observation; *SEA-ME-WE*, Southeast Asia–Middle East–Western Europe; *SRDRS*, submarine rescue diving and recompression system; *SS*, steamship; *SSK*, submarine hunter-killer; *SWISS*, shallow water intermediate search system; *TPL*, towed pinger locator; *TREC*, tethered remotely operated camera; *TROV*, tethered remotely operated vehicle; *USAF*, United States Air Force; *USCG*, United States Coast Guard; *USN*, United States Navy; *USNS*, United States Navy ship.

Vehicle	Location	Depth (fsw)	Year	Vessel	No. days	Company	Position	Mission
SCORPIO I	Pacific Ocean	2,000	1977	DB *Sampson*	60	Ocean Systems	Pilot/ROV tech.	Sea trials
SCORPIO I	**Formosa Straits (Taiwan)**	**1,400**	**1977**	***Deepsea I***	**30**	**Ocean Systems**	**Pilot/ROV tech.**	**Coral harvesting**
TROV S-4	Georgia Straits (Canada)	200	1978	Tugboat	14	Ocean Systems	Pilot/ROV tech.	Sea trials
Nekton Beta (manned)	Pacific Ocean	250	1978	MV *SeaMark*	14	General Oceanographics	Pilot/tech.	Training
Nekton Alpha (manned)	**Gulf of Mexico**	**400**	**1978**	**Workboat**	**14**	**General Oceanographics**	**Pilot/tech.**	**Pipeline inspection**
TROV S-4	Gulf of Mexico	250	1978	MV *Sea Rambler*	30	Ocean Systems	Pilot/ROV tech.	Pipeline inspection
TROV S-4	Gulf of Mexico	Standby	1978	DB 29	3	Ocean Systems	Pilot/ROV tech.	Pre-platform installation inspection

Vehicle	Location	Depth (fsw)	Year	Vessel	No. days	Company	Position	Mission
TREC	Gulf of Mexico	250	1978	Workboat	7	Ocean Systems	PM	Blowout preventer inspection
TROV S-4	Mediterranean Sea (Spain)	1,500	1978	MV *Kilsyth*	105	Ocean Systems	Sr. ROV tech.	Bottom survey
TREC	North Sea (Norway)	240	1979	MV *Ibis 7*	105	Solus Ocean Systems	Supervisor	Platform inspections
TROV S-4	North Sea (Norway)	240	1979	MV *Ibis 7*	60	Solus Ocean Systems	Supervisor	Platform cleaning and inspection
TREC	Gulf of Campeche (Mexico)	220	1980	DSV *Arctic Seal*	30	Underwater Services	Supervisor	Ixtoc I blowout investigation
TREC	Gulf of Mexico	450	1980	Workboat	30	Underwater Services	Consultant	Platform and riser survey
Towed camera sled	Gulf of Mexico (Florida)	300	1980	Workboat	14	Underwater Services	Consultant	Bottom survey
SCARABs I and II	N. Atlantic	3,000	1981	CS *Long Lines*	30	Ocean Search	Sr. ROV tech.	Sea trials
SCARAB II	Sea of Japan	1,500	1981	CS *KDD Maru*	30	Ocean Search	Sr. ROV tech.	Submarine cable inspection and repair
DART	North Sea (Norway)	450	1982	DP vessel	21	Nordex-Willco A.S.	Sr. ROV tech.	Concrete platform inspection
SCARAB II	N. Atlantic	250	1985	CCGS *John Cabot*	14	Eastport Int.	Sr. ROV tech.	Submarine cable inspection and repair
SCARAB II	**N. Atlantic (Ireland)**	**6,200**	**1985**	**CCGS *John Cabot***	**165**	**Eastport Int.**	**Sr. ROV tech.**	**Air India Flight 182 salvage**

Continued

Vehicle	Location	Depth (fsw)	Year	Vessel	No. days	Company	Position	Mission
Gemini	**N. Atlantic (Florida)**	**1,300**	**1986**	**DSV Stena Workhorse**	**75**	**Eastport Int.**	**Sr. ROV tech.**	**Space shuttle Challenger salvage**
ADS 620	**N. Atlantic (Florida)**	**80**	**1986**	**MV Independence**	**14**	**Eastport Int.**	**Sr. ROV tech.**	**Space shuttle Challenger salvage**
SCARAB II	N. Atlantic (Sable Island)	250	1986	CCGS John Cabot	21	Eastport Int.	Sr. ROV tech.	Submarine cable inspection and repair
Triton	Pacific Ocean (Japan)	1,500	1987	Tokai Salvage I	30	Eastport Int.	Sr. ROV tech.	PACAF F-16 aircraft recovery
Scorpi	Gulf of Mexico	480	1988	Dive ship	30	Eastport Int.	Sr. ROV tech.	Diving support, blowout investigation
Klein side-scan sonar	Wallops Island	80	1988	25-foot dive boat	2	Liberty Bell-7 Inc.	PM	Mercury capsule search
DOSS, Gemini 6000	N. Atlantic (Florida)	15,600	1992	MV Performer	14	Oceaneering Int.	PM	Liberty Bell 7 search
Magellan 725	N. Atlantic (Florida)	15,584	1993	Acoustic Pioneer	7	Oceaneering Int.	PM	Liberty Bell 7 search
Sea Plow V	**N. Atlantic (Halifax)**	**900**	**1994**	**CCGS John Cabot**	**14**	**Margus Telecom Int.**	**Field engineer**	**CANTAT-3 cable burial**
SCORPIO 2000	N. Atlantic (Halifax)	500	1994	CCGS John Cabot	21	Margus Telecom Int.	Field engineer	CANTAT-3 cable inspection
SCORPIO 2000	Mediterranean Sea (France)	1,700	1994	CS Raymond Croze	7	Margus Telecom Int.	Field engineer	Sea trials
SCORPIO 2000	Mediterranean Sea (Italy and France)	3,400	1994	CS Raymond Croze	23	Margus Telecom Int.	Field engineer	MAT2-D2 cable survey and burial

Vehicle	Location	Depth (fsw)	Year	Vessel	No. days	Company	Position	Mission
SCORPIO 2000	Mediterranean Sea (Italy)	3,400	1994	CS Raymond Croze	30	Margus Telecom Int.	Field engineer	Columbus-2 cable survey and burial
SCORPIO 2000	Mediterranean Sea (off Gibraltar)	3,040	1995	CS Raymond Croze	14	Margus Telecom Int.	Field engineer	Columbus-2 cable survey and burial
SCORPIO 2000	Mediterranean Sea	3,100	1995	CS Raymond Croze	16	Margus Telecom Int.	Field engineer	Finish Columbus-2 cable survey and burial
SCORPIO 2000	Mediterranean Sea	1,824	1995	CS Raymond Croze	18	Margus Telecom Int.	Field engineer	SEA-ME-WE 2 cable PLBO
SCORPIO 2000	Mediterranean Sea	3,000	1995	CS Raymond Croze	60	Margus Telecom Int.	Field engineer	SEA-ME-WE 2 cable PLBO
Sea Tractor 1	Grand Isle, LA	Surf	1996	Not applicable	42	Oceaneering Int.	Sr. ROV tech.	Beach trials
Phoenix II	S. China Sea	1,500	1996	MV Lammalco Snipe	75	Oceaneering Int.	Sr. ROV tech.	PLBO
USN MR-1	N. Atlantic (Long Island)	120	1996	USS Grasp	30	Oceaneering Int.	Asst. PM	TWA Flight 800 salvage
Deep Drone	N. Atlantic (Long Island)	120	1996	USS Grapple	21	Oceaneering Int.	PM	TWA Flight 800 salvage
Deep Drone	N. Atlantic (Long Island)	120	1996	MV Marion C	7	Oceaneering Int.	Asst. PM	TWA Flight 800 salvage
USN MR-3	N. Atlantic (Ocean City)	160	1996	USNS Apache	7	Oceaneering Int.	Asst. PM	FA-18 salvage
SWISS	E. Mediterranean (Israel)	5,000	1997	USNS Powhatan	7	Oceaneering Int.	Asst. PM	E-3B Viking search

Continued

Vehicle	Location	Depth (fsw)	Year	Vessel	No. days	Company	Position	Mission
Deep Drone	E. Mediterranean (Israel)	5,000	1997	USNS *Powhatan*	8	Oceaneering Int.	Asst. PM	E-3B recovery
CURV III	Cape Hatteras	230	1997	USNS *Grasp*	21	Oceaneering Int.	Asst. PM	HH-60 recovery
Sea Tractor	N. Atlantic	4,500	1997	USNS *Zeus*	16	Oceaneering Int.	Asst. PM	Sea trials
CRS II	English Channel	600	1997	USNS *Zeus*	45	Oceaneering Int.	Asst. PM	Cable burial
Magellan 725	N. Atlantic	12,600	1998	MV *Ocean Voyager*	25	Oceaneering Int.	Asst. PM	RMS *Titanic* operation
CRS II	English Channel	2,300	1998	USNS *Zeus*	32	Oceaneering Int.	Asst. PM	Cable burial
Ocean Explorer 6000	N. Atlantic	16,043	1999	MV *Needham Tide*	12	Liberty Bell-7 Inc.	PM	Liberty Bell 7 search
Ocean Discovery	N. Atlantic	16,043	1999	MV *Ocean Project*	18	Liberty Bell-7 Inc.	PM	Liberty Bell 7 recovery
SM-30	Philippine Sea	13,000	2000	MV *June T*	30	Liberty Bell-7 Inc.	PM	USS *Indianapolis* search
Remora 6000	Philippine Sea	13,000	2000	MV *Sea Eagle*	30	Liberty Bell-7 Inc.	PM	USS *Indianapolis* search
Mir I (manned)	Atlantic Ocean	16,043	2001	RV *Keldysh*	10	Atlantic Sands LLC	PM	Atlantic Target operation
CURV III	Atlantic Ocean	10,000	2002	MV *Powhatan*	21	Phoenix Int.	Asst. PM	CH-46 helicopter recovery
Deep Drone 8000	Gulf of Mexico	700	2002	MV *Ashley Candies*	10	Phoenix Int.	Asst. PM	T-39 trainer (2) search and recovery
Ocean Explorer 6000	S. Atlantic (Tierra del Fuego)	13,780	2003	MV *Secor Lenga*	21	Newport Explorations	PM	ARA *General Belgrano* search
Explorer	N. Atlantic (Nova Scotia)	200	2004	MV *Atlantic Hawk*	10	Phoenix Int.	Asst. PM	PLBO for International Telecom Group

Vehicle	Location	Depth (fsw)	Year	Vessel	No. days	Company	Position	Mission
SWISS, Magnum	Indian Ocean	3,700	2005	USS *Safeguard*	21	Phoenix Int.	Asst. PM	AV-8B Harrier search
CURV III	Atlantic Ocean	17,500	2005	MV *Jonathan Rozier*	14	Phoenix Int.	Asst. PM	Classified target object survey
CURV III	S. Pacific (off Fiji)	8,800	2007	MV *Seahorse Standard*	42	Phoenix Int.	Asst. PM	Black Hawk 221 search and recovery
CURV III	Indian Ocean (off Fremantle, Australia)	450	2007	MV *Seahorse Standard*	21	Phoenix Int.	Asst. PM	Remora SRDRS recovery
IT Explorer	N. Atlantic (off Newfoundland)	Not applicable	2007	CS *IT Intrepid*	14	Phoenix Int.	Asst. PM	ROV maintenance
Magnum 38	N. Atlantic	1,800	2007	USNS *Grasp*	14	Phoenix Int.	Asst. PM	E2-C Hawkeye search and recovery
SRDRS	San Diego, North Island	30	2007	Barge	35	Phoenix Int.	Asst. PM	SRDRS testing
SRDRS	San Diego, North Island	30	2008	Barge	25	Phoenix Int.	Asst. PM	SRDRS testing
Deep Drone 8000, SWISS	Indian Ocean (off Guam)	6,400	2008	USNS *Salvor*	14	Phoenix Int.	Asst. PM	EA6-B recovery
Deep Drone 8000, SWISS	Pacific Ocean (off Hawaii)	1,400	2008	USNS *Salvor*	14	Phoenix Int.	Asst. PM	SH-65 recovery (USCG)

Continued

Vehicle	Location	Depth (fsw)	Year	Vessel	No. days	Company	Position	Mission
Deep Drone 8000	Pacific Ocean (off San Clemente Island)	1,800	2009	USNS *Navajo*	7	Phoenix Int.	Asst. PM	AN/AQS-22 ALFS recovery, pressurized rescue module deep seat cleaning, acoustic array survey, submarine target vehicle inspection
Deep Drone 8000	Pacific Ocean (off San Diego)	1,800	2009	USNS *Salvor*	10	Phoenix Int.	Asst. PM	HH-60 helicopter search and recovery
TPL-40	**N. Atlantic (off Brazil)**	**13,000**	**2009**	**MV *Fairmount Glacier***	**35**	**Phoenix Int.**	**Asst. PM**	**Air France Flight 447 flight data recorder search**
TPL-40, SWISS, Deep Drone 8000	Pacific Ocean (off San Diego)	1,800	2009	USNS *Salvor*	10	Phoenix Int.	Asst. PM	USCG C130/USMC Cobra search and recovery
Fisher Defence SCORPIO, LR-5 submarine rescue vehicle	Indian Ocean (off Fremantle, Australia)	1,200	2010	MV *Seahorse Standard*	30	Phoenix Int., James Fisher Defence, Royal Australian Navy	Asst. PM	LR-5 manned submarine rescue vehicle sea trials, SCORPIO testing
TPL-21	**N. Arabian Sea**	**10,000**	**2010**	**USNS *Catawba***	**14**	**Phoenix Int.**	**Asst. PM**	**TPL search for EA6-B Hawkeye**
CURV 21	Pacific Ocean (NW of Hawaii)	18,500	2010	USNS *Navajo*	10	Phoenix Int.	Asst. PM	Classified package recovery
Deep Drone 8000, SWISS	Atlantic Ocean (off Norfolk, VA)	4,200	2011	USNS *Grapple*	7	Phoenix Int.	Asst. PM	ALFS dipping sonar search and recovery

Vehicle	Location	Depth (fsw)	Year	Vessel	No. days	Company	Position	Mission
SWISS, TRV-M	Gulf of Mexico (85 nautical miles south of Eglin AFB)	1,000	2011	MV *Marie Elise*	7	Phoenix Int.	PM	MALD missile search and recovery
Falcon	Pacific Ocean (off Santa Catalina Island)	476	2011	USNS *Salvor*	7	Phoenix Int.	PM	False seat installation
TPL-25	Gulf of Aden	9,000	2011	USNS *Catawba*	21	Phoenix Int.	Asst. PM	AV-8B Harrier search
Falcon	N. Atlantic	200	2011	Columbia barge	10	Phoenix Int.	PM	SSK target installation
Falcon	Gulf of Mexico	750	2011	Barge	5	Phoenix Int.	PM	SAT FADS diving support
Deep Drone	Spain	4,400	2012	USNS *Grasp*	45	Phoenix Int.	Asst. PM	Object recovery
Deep Drone 8000, SWISS	N. Atlantic	200	2012	USNS *Grasp*	10	Phoenix Int.	Asst. PM	Torpedo recovery
Magnum 38	N. Atlantic	1,500	2013	USNS *Grasp*	10	Phoenix Int.	Asst. PM	ALFS dipping sonar search and recovery
TPL-25	Taiwan Straits	900	2013	RV *Polaris*	10	Phoenix Int.	PM	ROCAF F-16 search
Seaeye Panther	Taiwan Straits	900	2013	RV *Polaris*	30	Phoenix Int.	PM	ROCAF F-16 recovery
TPL-25, Bluefin 21	S. Indian Ocean	15,000	2014	ADV *Ocean Shield*	60	Phoenix Int.	Asst. PM	Malaysia Airlines Flight 370 search
Deep Drone 8000, SWISS, TPL	N. Atlantic (off Panama City, FL)	550	2014	USNS *Grasp*	14	Phoenix Int.	Asst. PM	USAF F-16 search and recovery

Continued

Vehicle	Location	Depth (fsw)	Year	Vessel	No. days	Company	Position	Mission
Deep Drone 8000, SWISS, TPL	Atlantic Undersea Test and Evaluation Center	5,200	2014	USNS *Grasp*	3	Phoenix Int.	Asst. PM	MK-48 torpedo recovery
TPL, Orion, CURV 21	**N. Atlantic (off the Bahamas)**	**15,500**	**2015**	**USNS Apache**	**31**	**Phoenix Int.**	**Asst. PM**	**SS El Faro survey**
Iver 3-580-3037 AUV	Severn River, MD	20	2016	*Boston Whaler*	1	Phoenix Int.	PM	AUV trials
Iver 3-580-3037 AUV	Anacostia River	15	2016	Dock	1	Phoenix Int.	PM	USS *Barry* survey
Remus 6000	**Philippine Sea**	**15,000**	**2017**	**RV Petrel**	**45**	**Vulcan Inc.**	**Consultant**	**USS *Indianapolis* search**
F-15C survey	Sea of Japan	15,000	2018	RV *Petrel*	14	Phoenix Int.	PM	F-15C video inspection
TPL-25	**Philippine Sea**	**18,500**	**2018**	**MV Akatsuki**	**14**	**Phoenix Int.**	**PM**	**C-2A search**
Iver 3-580-3037 AUV	Alexandroupolis, Greece	30	2019	Dock	10	Phoenix Int.	PM	Harbor survey
Edgetech side-scan sonar	North Sea	120	2020	Unknown	14	Phoenix Int.	PM	F-15E search

Acronyms

ADS: advanced diving system

ARS: auxiliary rescue and salvage ship

AT&T: American Telephone and Telegraph

AUV: autonomous underwater vehicle

BEA: Bureau d'Enquêtes et d'Analyses

BOP: blowout preventer

C&W: Cable and Wireless

CCGS: Canadian Coast Guard ship

CS: cable ship

CURV: cable-controlled undersea recovery vehicle

DSV: diving support vessel

GPS: global positioning system

HiPAP: high precision acoustic positioning

LARS: launch and recovery system

LORAN: long-range navigation

LST: landing ship tank

MV: motor vessel

NTSB: National Transportation Safety Board

RCV: remotely controlled vehicle

ROV: remotely operated vehicle

RV: research vessel

SCARAB: submersible craft assisting repair and burial

SCORPIO: submersible craft for ocean repair, positioning, inspection, and
    observation

SS: steam ship

T-ATF: auxiliary tug fleet

TPL: towed pinger locator

TREC: tethered remotely operated camera

TROV: tethered remotely operated vehicle

USAF: United States Air Force
USCG: United States Coast Guard
USN: United States Navy
USNS: United States Navy ship
USS: United States ship

Sources

Anderson, M. A. "Memorandum, Rescue and Search for Survivors of USS *Indianapolis* (CA 35) and Recovery, Identification, and Burial of Bodies, File: P1/L11-1," 15 August 1945.

"Atlas of Pilot Charts of the North Pacific Ocean," DOD, DMA, NVPUB108, 3rd ed., 1994.

Baker, Richard Walter. "Ancestors of Elswick Newport," 20 February 2010.

Claytor, W. G. Jr., Lt. Commander, USN, Commanding Officer. "Memorandum, Report on Rescue of Survivors of USS *Indianapolis* (CA 35) August 2–4, 1945, USS *Cecil J. Doyle* (DE 368)," 25 May 2010.

"Command Investigation into the Facts and Circumstances Surrounding the VAW-121 Aviation Mishap and Fatality That Occurred on 31 March 2010 in the Gulf of Oman," Commander, Carrier Strike Group Eight, 25 May 2010.

Decklog, USS *Alvin C. Cockrell* (DE 366), 1–8 August 1945.

Decklog, USS *Aylwin* (DD 355), 1–8 August 1945.

Decklog, USS *Bassett* (APD 73), 1–4 August 1945.

Decklog, USS *Cecil J. Doyle* (DE 368), 1–9 August 1945.

Decklog, USS *Dufilho* (DE 423), 1–8 August 1945.

Decklog, USS *French* (DE 367), 4–10 August 1945.

Decklog, USS *Helm* (DD 388), 1–7 August 1945.

Decklog, USS *Madison* (DD 425), 1–6 August 1945.

Decklog, USS *Ralph Talbot* (DD 390), 1–6 August 1945.

Decklog, USS *Ringness* (APD 100), 2–4 August 1945.

Finke, John, Phoenix International Holdings Inc., interview with C. Newport, 13 January 2023.

Furman, J. R. "Memorandum, Search Operations of USS *Register* (APD 92) for Survivors of USS *Indianapolis*, File: APD92/A16-3," 8 August 1945.

Gwinn, W. G., Lt. USNR, Lt. R. A. Marks, USNR, and Lt. Cmdr. G. C. Atteberry, USN, transcript of interview, 6 August 1945.

Kurzman, Dan. *Fatal Voyage: The Sinking of the USS* Indianapolis. Atheneum, 1990.

Langford, M. S., Lt. USNR, Air Combat Intelligence Officer. "Sighting of Survivors of the U.S.S. *Indianapolis*; Participation in Air-Sea Rescue and Subsequent Search for Bodies and Debris 2–7 August 1945," 9 August 1945.

Leopold, George. Did Static Electricity—Not Gus Grissom—Blow the Hatch of the *Liberty Bell 7* Spacecraft? *Astronomy*, 21 July 2021. https://www.astronomy.com/space-exploration/did-static-electricity-not-gus-grissom-blow-the-hatch-of-the-liberty-bell-7-spacecraft/.

Lindbergh, Charles A. *Autobiography of Values*. Harcourt Brace Jovanovich, 1976.

Lindbergh, Charles A. *The Wartime Journals of Charles A. Lindbergh*. Harcourt Brace Jovanovich, 1970.

"Marine Accident Report, Sinking of US Cargo Vessel SS *El Faro*, Atlantic Ocean, Northeast of Acklins and Crooked Islands, Bahamas," Accident Report NTSB/MAR-17/01, 1 October 2015.

"Memorandum, USS *Indianapolis* (CA 35)—Loss of McVay, Charles B. III," undated.

Navigational chart, Philippine Islands to Bismarck Archipelago, Hydrographic Department, M.S.A., Japan, Nima Ref. No. WOPGN507.

Newport, Curt. *Lost Spacecraft: The Search for* Liberty Bell 7. Apogee Books, 2002.

Newport, Elswick, Capt., Award of the Distinguished Flying Cross, Eighth United States Army, 2 August 1953.

Newport, Elswick, Lt. Col., Award of the Air Medal for Heroism, with "V" Device, United States Army Vietnam, 13 April 1966.

Radio dispatch, USS *Ringness* (APD 100), COMINCH F-32 Combatant Damage Card, 2 August 1945.

"Record of Proceedings of a Court of Inquiry, Convened at Headquarters, Commander Marianas, Guam, Commander in Chief, United States Pacific Fleet and Pacific Ocean Areas," 13 August 1945.

"Record of Proceedings of a General Court Martial, Case of Charles B. McVay, 3rd, US Navy Judge Advocate General (JAG), Volumes I, II, and III," 3 December 1945.

"Rescue of Survivors of USS *Indianapolis*, 2–6 August 1945, USS *Ringness* (APD 100)," Action Report, 6 August 1945.

"Statistics of Ocean Current in One Degree Mesh, Months July–August," Japan Oceanographic Data Center, 12 January 1998.

"STS 51-L *Challenger* Search and Salvage Operation," Department of the Navy, Naval Sea Systems Command, 10 October 1986.

Twible, Harlan M., Ensign, USS *Indianapolis*, USN (Ret), correspondence with C. Newport, November–December 1999.

"U.S. Navy Marine Climatic Atlas of the World," Ver. 1.1, August 1995.

"U.S. Navy Salvage Report: TWA Flight 800," S0300-BZ-RPT-010, 0910-LP-015-6130, Commander, Naval Sea Systems Command, May 1998.

Wendt, Guenter F. *The Unbroken Chain*, Apogee Books, 2001.

About the Author

Curt Newport retired in 2022 from Phoenix International Holdings after nearly fifty years in the subsea industry. During his career, he operated Canadian, US, British, and Norwegian vehicles on an international basis, ranging from the Arctic to the Roaring Forties, the Persian Gulf, the North Sea, and both sides of the equator in the Atlantic and Pacific Oceans, and participated in more than 150 undersea operations in water depths down to 5,500 meters (over 3.4 miles).

Operations Newport supported include the salvage of Air India Flight 182, the space shuttle *Challenger*, the recovery of Gus Grissom's *Liberty Bell 7*, TWA Flight 800, the broadcast of live images from the RMS *Titanic*, Air France Flight 447, Malaysia Airlines Flight 370, the SS *El Faro* investigation, and the USS *Indianapolis*, as well as many classified missions involving the loss of military aircraft.

Along with oceanographic engineer Dr. Toby Schneider, Newport developed an autonomous pinger locator payload that successfully localized a simulated flight data recorder in Buzzards Bay, Massachusetts, using an Iver 3-580 autonomous underwater vehicle.

Newport's exploration interests are not restricted to the deep ocean. In 2004, he successfully probed the stratosphere using his Proteus 6 experimental rocket in the Black Rock Desert, Nevada. The 200-pound vehicle reached a velocity of Mach 3 and an altitude of 14 miles while recording HD video and transmitting GPS data via a radio frequency link to his ground station. He has also participated in cave explorations, helping to map previously unexplored passages in Simmons-Mingo Cave (West Virginia) and Cave of the Winds (Colorado), as well as noncommercial sections of Luray Caverns (Virginia).

Newport has lectured on behalf of the Smithsonian's National Air and Space Museum, Kansas Aviation Museum, Children's Museum of Indianapolis, Boston's Museum of Science, St. Louis Science Center, and the Kansas Cosmosphere and Space Center and has been interviewed by various media organizations, such as NBC's *Today Show* and *Evening News*, ABC's *Good Morning America*, CNN, *CBS Morning News*, *USA Today*, *Florida Today*, the British Broadcasting Company, the Canadian Broadcasting Company, National Public Radio, CBS National Radio, Reuters, the History Channel, and the Associated Press.

He currently resides in Mt. Vernon, Virginia, where he spends his time writing, shooting antique black powder rifles, and restoring his 1959 Austin Healey Sprite.